Lectures in Mathematics
ETH Zürich
Department of Mathematics
Research Institute of Mathematics

Managing Editor:
Helmut Hofer

Mark Freidlin

Markov Processes and Differential Equations: Asymptotic Problems

Springer Basel AG

Author's Address:

Mark Freidlin
Department of Mathematics
University of Maryland
College Park
MD 20742
USA

Mathematics Subject Classification (1991): 60J60, 60HXX, 58G32, 35B40, 35K55

A CIP catalogue record for this book is available from the Library of Congress,
Washington D.C., USA

Deutsche Bibliothek Cataloging-in-Publication Data
Frejdlin, Mark I.:
Markov processes and differential equations : asymptotic
problems / Mark Freidlin. - Basel ; Boston ; Berlin :
Birkhäuser, 1996
  (Lectures in mathematics)
  ISBN 978-3-7643-5392-6   ISBN 978-3-0348-9191-2 (eBook)
  DOI 10.1007/978-3-0348-9191-2

© 1996 Springer Basel AG
Originally published by Birkhäuser Verlag, P.O. Box 133, CH-4010 Basel, Switzerland in 1996

ISBN 978-3-7643-5392-6

9 8 7 6 5 4 3 2 1

Contents

Preface

We study in these lectures a number of asymptotic problems arising in the theory of stochastic processes and in the theory of partial differential equations. Due to the close connection between the second order differential operators with a non-negative characteristic form on the one hand and Markov processes on the other, many problems for PDE's can be reformulated as problems for corresponding stochastic processes and vice versa. Most of the problems considered here have their origin in PDE's theory, although the statements, as a rule, are natural from a probabilistic point of view as well.

We consider four groups of problems:

The Dirichlet problem with a small parameter in higher derivatives for differential equations and systems;

The averaging principle for stochastic processes and PDE's;

Homogenization in PDE's and in stochastic processes;

Wave front propagation for semilinear differential equations and systems.

From the probabilistic point of view, the first two topics concern random perturbations of dynamical systems. The third topic, homogenization, is a natural problem for stochastic processes as well as for PDE's. Wave fronts in semilinear PDE's is one of the interesting and important for applications examples of pattern formation in the reaction-diffusion equations (RDE's). Our machinery is mostly probabilistic: limit theorems of law-of-large numbers type, central-limit-theorem type, and of large-deviation type are our main tools.

Most attention is paid to relatively recent results: boundary problems with a small parameter in the higher derivatives for a class of PDE systems, the averaging principle for randomly perturbed dynamical systems with conservation laws, RDE's in narrow tubes, large scale approximation for RDE's, general conditions for homogenization of one-dimensional processes, homogenization for the large deviations. But we include also some older results which are the basis for the new development.

We did not try to consider the problems in their most general setup and prefered to demonstrate interesting effects in the simplest situation where they still appear.

As a rule, we do not go into detail in the proofs if the detailed proofs are available in the literature. We restrict ourselves to the explanation of the main ideas in the simplest situation and give the references.

These lectures were delivered as Rudolph Lipschitz Lectures at Bonn University in the spring semester of 1994 and later, in a slightly extended form at ETH, Zurich. I am sincerely grateful to my colleagues at Bonn University and at ETH. I would especially like to thank Hans Föllmer and Alain - Sol Sznitman for friendly and interesting discussions of many topics related and not related to these lectures.

I would like to also thank Dr. James Dunyak, Errol Rowe and Anja Voss who helped me in the preparation of the manuscript.

The work was supported in part by ARO Grant DAAL03-92-G0219 and NSF Grant DMS-9504177.

College Park,
August 1995

1 Stochastic Processes Defined by ODE's

A stochastic process in the time interval $[0, \infty)$ is defined as a family of random variables $X_t(\omega)$, $t \geq 0$, $\omega \in \Omega$, on a measurable space $(\Omega, \mathcal{F}, \mathcal{P})$. To describe the probability structure of the process, one should define the family of distributions of $(X_{t_1}, \ldots, X_{t_n})$ for any integer n and any $0 \leq t_1 < t_2 < \ldots < t_n$. This family of finite dimensional distributions is, in general, a rather bulky subject. Therefore, as a rule, special classes of stochastic processes are considered for which such a description can be reduced to more convenient characteristics. First, we consider the basic processes which have a simple statistical structure. Then we consider relatively simple and explicitly defined transformations of these basic processes. The Wiener process, Poisson process, and continuous time Markov chains with finite number of states will be our basic processes. We assume that the main properties of these processes are known. Actually, all classes of continuous time stochastic processes, allowing deep enough theory, can be constructed from these basic processes using relatively simple transformations.

Let W_t be the one-dimensional Wiener process. This means that $W_t = W_t(\omega)$, $t \geq 0$, $\omega \in \Omega$, is a continuous with probability one, mean zero Gaussian process such that $EW_sW_t = s \wedge t$. (E is the sign of expectation.) One can prove that such a process exists and has independent increments. The collection $(W_t^{(1)}, \ldots, W_t^{(r)}) = W_t$ of r independent Wiener processes is called a r-dimensional Wiener process. See, for example, [F1, §1.2] for properties of the Wiener process. It follows from the definition that, for any $t > 0$, the r-dimensional random variable W_t has a density

$$p(t, y) = \frac{1}{(2\pi t)^{r/2}} \exp\left\{-\frac{|y|^2}{2t}\right\}.$$

Using this fact, it is easy to check that for any bounded continuous $g(x)$, $x \in \mathbb{R}^r$, the function

$$u(t, x) = Eg(x + W_t)$$

is the solution of the Cauchy problem

$$
\begin{aligned}
\frac{\partial u(t, x)}{\partial t} &= \Delta u(t, x), \ t > 0, \ x \in \mathbb{R}, \\
u(0, x) &= g(x).
\end{aligned}
$$

This is a manifestation of the close connection between the Wiener process and the simplest heat equation.

Consider now a general second order differential operator with a non-negative characteristic form:

$$L = \frac{1}{2} \sum_{i,j=1}^{r} a^{ij}(x) \frac{\partial^2}{\partial x^i \partial x^j} + \sum_{i=1}^{r} b^i(x) \frac{\partial}{\partial x^i}, \quad \sum_{i,j} a^{ij}(x) \lambda_i \lambda_j \geq 0, \ x \in \mathbb{R}^r.$$

Since the matrix $(a^{ij}(x))$ is non-negatively defined, a matrix $\sigma(x)$ exists such that $\sigma(x)\sigma^*(x) = (a^{ij}(x))$. For example, one can take as $\sigma(x)$ the symmetric square root of $(a^{ij}(x))$. We assume that the matrix $\sigma(x)$, $x \in \mathrm{R}^r$, can be chosen so that it is Lipschitz continuous. If the matrix $(a^{ij}(x))$ is non-degenerate then the existence of a Lipschitz continuous $\sigma(x)$ follows from the Lipschitz continuity of $(a^{ij}(x))$. If we allow degeneration of $(a^{ij}(x))$ then a Lipschitz continuous $\sigma(x)$ exists if, at least, the entries $a^{ij}(x)$ have bounded second derivatives (see §3.2 in [F6]). We assume that the coefficients $b^i(x)$ are Lipschitz continuous as well.

Consider the differential equation

$$\dot{X}_t^x = b(X_t^x) + \sigma(X_t^x)\dot{W}_t, \ \ X_0^x = x \in \mathrm{R}^r, \ t \geq 0, \tag{1.1}$$

where W_t is the Wiener process in R^r and $\dot{W}_t$ is the white noise process. If $\sigma(x) = \sigma$ is independent of x, the existence and uniqueness of the solution for this equation can be proved exactly in the same way as for ordinary differential equations for any fixed continuous trajectory W_t, $t \geq 0$. Define

$$\phi_0(t) \equiv x, \ \ \phi_{n+1}(t) = x + \int_0^t b(\phi_n(s))ds + \sigma W_t.$$

Since we assumed Lipschitz continuity of the coefficients these approximations converge uniformly in any finite time interval.

If $\sigma(x)$ depends on x, (1.1) should be understood as an Itô equation (See, for example, [F1], §1.3). The drift $b = (b^1, \dots, b^r)$ and the matrix σ in (1.1) can also depend on t.

Equation (1.1) defines the transformation of the Wiener process W_t in a new process X_t^x (actually, a family of processes X_t^x depending on the parameter $x \in \mathrm{R}^r$). This family is closely connected with the operator L.

It follows from (1.1), that

$$X_t^x - X_{t_0}^x = \int_{t_0}^t \sigma(X_s^x)\,dW_s + \int_{t_0}^t b(X_s^x)\,ds, \ t \geq t_0 \geq 0.$$

Taking into account that the Wiener process has independent increments and that equation (1.1) has a unique solution, one can check that the process X_t^x has the Markov property (see [D], [F6, §1.4]).

The collection X_t^x of the solutions of (1.1) for various $x \in \mathrm{R}^r$, together with the probabilistic measure P (on the sample space for the basic Wiener process $W.$), is called the Markov family corresponding to the operator L.

The set of probabilistic measures $P_x(\cdot)$, on the space of continuous functions on $[0, \infty)$ with values in R^r, defined by the equality

$$P_x(A) = P\{X_\cdot^x \in A\},$$

for Borel sets A, is called the Markov process corresponding to L.

In general, the Markov process in the phase space $\mathcal{E}$ is a family of probabilistic measures P_x, $x \in \mathcal{E}$, in the space of functions $[0,\infty) \to \mathcal{E}$, satisfying certain relations [D]. It is sometimes convenient to include in the notations the trajectories. For example, (X_t, P_x) is the Markov process corresponding to L, if P_x are the measures defined above, and $X.$ is a continuous function on $[0,\infty)$ with values in R^r. We will see other examples later.

We will write E_x for the expectation sign with respect to the measure P_x; the sign E means expectation with respect to the measure P. It is clear that we can go from one to the other: for example, $E_x f(X_t) = E f(X_t^x)$.

Let $u(t, x)$ be a continuously differentiable once in t and twice in $x \in \mathrm{R}^r$ function. The following Itô formula holds (see, for example, [F6], §1.3):

$$
u\left(t - t_2, X_{t_2}^x\right) - u\left(t - t_1, X_{t_1}^x\right) = -\int_{t_1}^{t_2} \frac{\partial u\left(t - s, X_s^x\right)}{\partial t}\, ds
$$
$$
+ \int_{t_1}^{t_2} L u\left(t - s, X_s^x\right) ds + \int_{t_1}^{t_2} \nabla u\left(t - s, X_s^x\right) \cdot \sigma\left(X_s^x\right) dW_s, \tag{1.2}
$$
$$
0 \le t \le t_1 \le t_2, \ x \in \mathrm{R}^r.
$$

The last term on the right side of (1.2) is a stochastic Itô integral; $\nabla u(t, x)$ is the gradient in $x \in \mathrm{R}^r$.

Recall that the Itô integral $\int_0^t f(s, \omega) dW_s$ is defined at least for the "independent of future" functions $f(s, \omega)$, $0 \le s \le t$, $\omega \in \Omega$, such that

$$
\int_0^t E|f^2(s, \omega)|\, ds < \infty.
$$

Independence of future means that $f(s, \omega)$ is independent of the behavior of the basic Wiener process $W.$ after time s. For such $f_1(s, \omega)$, $f_2(s, \omega)$ the Itô integral, besides the standard properties, satisfies the two relations:

$$
E \int_0^t f_i(s, \omega)\, dW_s \ = \ 0
$$
$$
E\left(\int_0^t f_1(s, \omega)\, dW_s \cdot \int_0^t f_2(s, \omega)\, dW_s\right) \ = \ \int_0^t E f_1(s, \omega) f_2(s, \omega) ds
$$

(See, for example, [F6]).

One can connect with any Markov process a semigroup of operators T_t, $t > 0$, acting in the space of bounded measurable functions on the phase space:

$$
(T_t f)(x) = E_x f(X_t) = E f(X_t^x). \tag{1.3}
$$

The semigroup property of the operators T_t,

$$T_{s+t} = T_s T_t = T_t T_s,$$

is equivalent to the Markov property of the process (X_t, P_x) (family (X_t^x, P)). The generator A of the semigroup T_t (of the process (X_t, P_x)) is defined as follows:

$$Af = \lim_{t \downarrow 0} \frac{T_t f - f}{t} \text{ (limit in the norm)},$$

Using the Itô formula, one can easily check that the generator A of the semigroup T_t is defined for smooth enough functions $f(x)$, $x \in \mathbb{R}^r$; and for such functions, $Af(x) = Lf(x)$. As in the case of solutions of ordinary differential equations, the solutions of the Itô equations with Lipschitz continous coefficients depend continuously on the initial conditions. This yields that the semigroup T_t maps the space of bounded continuous functions on $\mathbb{R}^r$ into itself. This property, together with the continuity of the trajectories, provides the strong Markov property of the process (X_t, P_x) definded by (1.1) (see [D], [F6]).

The stochastic differential equation is not the only way to construct the process corresponding to the operator L. The formula (1.2) implies that for any $x \in \mathbb{R}^r$ and any smooth function $f(x)$, the process

$$Z_t = f(X_t) - \int_0^t Lf(X_s)\, ds$$

is a martingale with respect to the measure P_x and the family of σ-fields $\mathcal{F}_t$ generated by the process X_s for $s \leq t$. This property may be taken as the definition of the Markov process corresponding to the operator L (see [SV], [EK]). Such an approach allows us to prove existence and uniqueness of the process under very weak assumptions on the coefficients of the operator L. This approach is useful when proving limit theorems for stochastic processes.

We will consider now the probabilistic representations for solutions of some initial-boundary problems related to the operator L([F6], Ch. 2). Assume for brevity that L is a non-degenerate elliptic operator, and let G be a bounded domain on $\mathbb{R}^r$ with a smooth enough boundary.

Consider the Dirichlet problem

$$Lu(x) = 0, \ x \in G, \ u(x) \mid_{\partial G} = \psi(x),$$

where $\psi(x)$ is continuous. Then the unique solution of this problem has the representation:

$$u(x) = E_x \psi(X_\tau). \tag{1.4}$$

Here (X_t, P_x) is the Markov process corresponding to L, and τ is the first exit time from the domain G, i.e., $\tau = \min\{t : X_t \notin G\}$.

To prove (1.4) assume first that $\psi(x)$ and ∂G are smooth enough so that a C^2-class function $V(x)$, $x \in \mathbb{R}^r$, exists coinciding with $u(x)$ on G. Applying the Itô formula, we have:

$$V(X_t^x) = V(x) + \int_0^t \nabla V(X_s^x) \cdot \sigma(X_s^x)\, dW_s + \int_0^t LV(X_s^x)\, ds.$$

It is easy to check that $E_x \tau < \infty$, $x \in G$. Then, putting $t = \tau$ in the last formula and taking into account that the expectation of the stochastic integral is equal to zero and $LV(X_s^x) = 0$ for $0 \le s < \tau$, we derive (1.4). The case of non-smooth $\psi(x)$ can be reduced to the smooth one by approximation (see [F6]).

The exit time τ is an example of a so called Markov time-independent of future non-negative random variable. The exact definition and properties are found in standard textbooks for Markov processes (see also [F6], §1.4). If τ is a finite with probability one Markov time then the behavior of the process (X_t, P_x) for $t > \tau$ given X_τ is independent of the process X_s for $s < \tau$. This is a form of the strong Markov property mentioned above.

Consider now the initial boundary value problem:

$$\frac{\partial u(t,x)}{\partial t} = Lu(t,x) + c(t,x)u(t,x), \ \ t > 0, \ x \in G \subset \mathbb{R}^r, \tag{1.5}$$

$$u(0,x) = g(x), \ \ u(t,x)\,|_{x \in G, \ t>0} = \psi(x).$$

Here $c(t,x), g(x)$ and $\psi(x)$ are continuous bounded functions. The domain G has a smooth enough boundary but can be unbounded.

Using the Itô formula, one can derive the representation for the solution of problem (1.5) which is called the Feynman-Kac formula:

$$u(t,x) = E_x\, g(X_t)\, e^{\int_0^t c(t-s, X_s)ds} \cdot \chi_{\tau > t} + E_x \psi(X_\tau)\, e^{\int_0^\tau c(t-s, X_s)ds} \cdot \chi_{\tau \le t}. \tag{1.6}$$

Here $\chi_{\tau > t}$ is the indicator function of the set $\{\tau > t\}$ in $C_{0,\infty}$ and $\chi_{\tau \le t} = 1 - \chi_{\tau > t}$. In particular, if $G = \mathbb{R}^r$, the second term in (1.6) disappears, and we have the representation of the solution of the Cauchy problem.

Consider together with the process X_t^x defined by (1.1) the process $\tilde{X}_t^x$:

$$\dot{\tilde{X}}_t^x = b(\tilde{X}_t^x) + \sigma(\tilde{X}_t^x)\dot{W}_t + \tilde{b}(t, \tilde{X}_t^x), \ \ \tilde{X}_0^x = x, \tag{1.7}$$

for some bounded $\tilde{b}(x)$. Denote by $\tilde{P}_x$ the measure in $C_{0,T}$, $0 < T < \infty$, corresponding to the process $\tilde{X}_t^x$. It turns out that the measures P_x, corresponding to X_t^x, and $\tilde{P}_x$, under some mild conditions, are absolutely continuous. Namely, assume that the system of linear equations

$$\sigma(x)\varphi(t,x) = \tilde{b}(t,x)$$

is solvable for any $x \in \mathbb{R}^r$, $0 \le t \le T$, and that the solution $\varphi(t,x)$ is bounded uniformly in $x \in \mathbb{R}^r$, $0 \le t \le T$. (We do not assume that the matrix $a(x) = \sigma(x)\sigma^*(x)$ is non-degenerate. If $\det(a(x)) \ne 0$, the linear system is, of course, always solvable). Then the measures $\tilde{P}_x$ and P_x in C_{0T} are absolutely continuous with respect to each other and (see, example, [F1] §1.5)

$$\frac{d\tilde{P}}{dP}(h.) = \exp\left\{ \int_0^T \varphi(s, h_s) \cdot dW_s - \frac{1}{2}\int_0^t |\varphi(s, h_s)|^2 \, ds \right\}. \qquad (1.8)$$

Probabilistic representation for the solutions of linear problems can be used for studying related non-linear problems ([F6], Ch. 5-7). Consider, for example, the Cauchy problem for a quasilinear parabolic equation:

$$\frac{\partial u(t,x)}{\partial t} = \frac{1}{2}\sum_{i,j=1}^r a^{ij}(x,u)\frac{\partial^2 u}{\partial x^i \, \partial x^j} + \sum_{i=1}^r b^i(x,u)\frac{\partial u}{\partial x^i} + c(x,u)u, \qquad (1.9)$$

$$t > 0, \ x \in \mathbb{R}^r, \ u(0,x) = g(x).$$

If (1.9) is solvable then $u(t,x)$ together with $X_t^{x,t}$ satisfy the equations:

$$\dot{X}_s^{x,t} = \sigma\left(X_s^{x,t}, u\left(t - s, X_s^{x,t}\right)\right)\dot{W}_s + b\left(X_s^{x,t}, u\left(t - s, X_x^{x,t}\right)\right), \ X_0^{x,t} = x,$$

$$u(t,x) = Eg\left(X_t^{x,t}\right)\exp\left\{ \int_0^t c\left(X_s^{x,t}, u\left(t - s, X_s^{x,t}\right)\right) ds \right\},$$

$$0 \le s \le t < \infty, \ x \in \mathbb{R}^r, \ \sigma(x,u)\sigma^*(x,u) = \left(a^{ij}(x,u)\right),$$

$$b(x,u) = (b^1(x,u), \dots, b^r(x,u)).$$

$$(1.10)$$

This follows from (1.6). If the coefficients $a^{ij}(x,u)$ depend only on x, and the linear system $\sigma(x)\varphi(x,u) = b(x,u)$ is solvable for any $x \in \mathbb{R}^r$ and $|u| < \infty$, the system (1.10) can be reduced to a triangle one: denoting by Y_t^x the solution of the "shortened" system

$$\dot{Y}_t^x = \sigma(Y_t^x)\dot{W}_t, \ Y_0^x = x, \qquad (1.11)$$

and using (1.8) we conclude that the function $u(t,x)$ satisfies the equation

$$u(t,x) = E_x g(Y_t^x)\exp\left\{ \int_0^x \varphi(Y_s^x, u(t-s, Y_s^x)) \cdot dW_s + \right.$$

$$\left. + \int_0^t \left[c(Y_s^x, u(t-s, Y_s^x)) - \frac{1}{2}|\varphi(Y_s^x, u(t-s, Y_s^x))|^2\right] ds \right\}. \qquad (1.12)$$

Equations (1.11) and (1.12) form the triangle system for $(Y_t^x, u(t,x))$. One can use it, for example, to prove existence and uniqueness of the solution of problem (1.9) as well as for studying the smoothness of the solution. Such an approach is especially useful when the equation degenerates ([F6], Ch. 5). We will use (1.12) in the case $\varphi \equiv 0$ later when studying the wavefront propagation for reaction-diffusion equations (RDE).

By a reaction-diffusion equation system we understand one equation or a system of the form

$$\frac{\partial u_k(t,x)}{\partial t} = \frac{1}{2} \sum_{i,j=1}^{r} a_k^{ij}(x) \frac{\partial^2 u_k}{\partial x^i \partial x^j} + \sum_{i=1}^{r} b_k^i(x) \frac{\partial u_k}{\partial x^i} + f_k(x; u_1, \ldots, u_k) =$$

$$= L_k u_k + f_k(x; u), \tag{1.13}$$

$$u_k(0, x) = g_k(x), \quad k = 1, \cdots n; \quad x \in \mathbb{R}^r, \; t > 0.$$

The operators $L_k, k = 1, \ldots, n$, are assumed to be elliptic (at least $\sum\limits_{i,j=1}^{r} a_k^{ij}(x)\lambda^i\lambda^j$ ≥ 0 for $x \in \mathbb{R}^r$, $k = 1, \ldots, n$) with bounded, sufficiently regular coefficients. If equations (1.13) are considered not in the whole space $\mathbb{R}^r$ but in a domain $G \subset \mathbb{R}^r$, some boundary conditions on ∂G should be added.

Equations (1.13) describe the time evolution of a system consisting of n types of particles. The operators L_k govern the motion of the particles. The nonlinear terms describe mutual transmutations of the particles.

If $n = 1$ and $f_1 = f(x, u) = c(x, u)u$ and (X_t, P_x) is the Markov process corresponding to the operator $L_1 = L$, then (1.12) gives the following integral equation in the functional space for the unknown function $u_1(t, x) = u(t, x)$:

$$u(t, x) = E_x \, g(X_t) \exp \left\{ \int_0^t c\left(X_s, u(t - s, X_s)\right) ds \right\}. \tag{1.14}$$

To write down a similar equation for $u_k(t, x)$ in the case of systems $(n > 1)$ one should first consider linear systems of the type (1.13):

$$\frac{\partial u_k(t,x)}{\partial t} = L_k u_k + \sum_{i=1}^{n} c_{ki}(x)(u_i - u_k) + c_k(t, x) \cdot u_k, \tag{1.15}$$

$$u_k(0, x) = g_k(x), \; x \in \mathbb{R}^r, \; t > 0, \; k = 1, \ldots, n.$$

Assume that $c_{kj}(x) \geq 0$ for $j \neq k$. Let $\left(a_k^{ij}(x)\right) = \sigma_k(x)\sigma_k^*(x)$, $b_k(x) = (b_k^1(x), \ldots$ $\ldots, b_k^r(x))$, $k = 1, \ldots, n$. Consider the Markov process $(X_t, \nu_t; P_{x,k})$ in the phase space $\mathbb{R}^r \times \{1, \ldots, n\}$:

$$\dot{X}_t^{x,k} = \sigma_{\nu_t^{x,k}}(X_t^{x,k})\dot{W}_t + b_{\nu_t^{x,k}}(X_t^{x,k}), \; X_0^{x,k} = x, \; \nu_0^{x,k} = k,$$

$$P\left\{\nu_{t+\Delta}^{x,k} = j \; \Big| \; X_t^{x,k} = x, \; \nu_t^{x,k} = i\right\} = c_{ij}(x)\Delta + o(\Delta), \; \Delta \downarrow 0. \tag{1.16}$$

We assume that $\nu_t^{x,k}$ is right continuous with probability 1. Such a process $(X_t, \nu_t; P_{x,k})$ exists, at least, if the coefficients $\sigma_k(x), b_k(x), c_{ij}(x)$ are bounded and Lipschitz continuous. Note that in the case of constant coefficients $c_{kj}(x) = c_{kj}$, the component $\nu_t^{k,x} = \nu_t^k$ is defined independently of $X_t^{x,k}$. The process ν_t^k in that case is the continuous time Markov chain in the phase space $\{1, \ldots, n\}$ with the intensities c_{kj}, $\nu_0^k = k$. Then the first of equations (1.16) defines the component $X_t^{x,k}$ for a given trajectory ν_t^k. One can check [EF1], that the solution of problems (1.15) has the following representation

$$u_k(t,x) = E_{x,k} g_{\nu_t}(X_t) \exp \left\{ \int_0^t c_{\nu_s}(t-s, X_s) \, ds \right\}. \tag{1.17}$$

As before, the subscript in the expectation sign means the starting point for the trajectory (X_t, ν_t).

Representation (1.17) holds for the solution of problem (1.15) if $c_{ij}(x) \geq 0$, $i \neq j$. Only in this case one can define the processes $(X_t, \nu_t, P_{x,k})$. If we consider physical problems leading to RDE's, the functions $u_k(t,x)$ have the sense of density functions. Thus $u_k(t,x)$ should always be nonnegative if the initial conditions are nonnegative. One can show that this property is fulfilled for a linear system only if $c_{ij}(x) \geq 0$ for $i \neq j$. But if we consider nonlinear RDE systems and $c_{ij} = c_{ij}(x,u)$, the positiveness condition can be satisfied even if $c_{ij}(x,u)$ are negative for some u. Thus, to incorporate the general case, we need a probabilistic representation for the solutions of system (1.15) in the case that $c_{ij}(x)$ is of any sign. One can use the following trick. Consider an auxiliary process $(\widehat{X}_t, \widehat{\nu}_t; \widehat{P}_{x,k})$ defined by equations (1.16) with $c_{ij}(x) \equiv 1$, $i, j = 1, \ldots, n$, $x \in \mathbb{R}^r$. The process $(X_t, \nu_t; P_{x,k})$ corresponding to system (1.15) with nonnegative $c_{ij}(x)$ is absolutely continuous with respect to the auxiliary process $(\widehat{X}_t, \widehat{\nu}_t; \widehat{P}_{x,k})$ for any finite time interval [0,T], and

$$\frac{dP_{x,k}}{d\widehat{P}_{x,k}}(X., \nu.) =$$

$$\prod_{i=0}^{n(T)-1} c_{\nu_{\eta_i}, \nu_{\eta_{i+1}}}(X_{\eta_i}) \exp \left\{ -\sum_{i=0}^{n(T)} \int_{\eta_i}^{T \wedge \eta_{i+1}} [\hat{c}_{\nu_i}(X_s) - n + 1] \, ds \right\}, \tag{1.18}$$

where $\hat{c}_k(x) = \sum_{\substack{i=1 \\ i \neq k}}^{r} c_{ki}(x)$, and $\{\eta_i\}$ is a sequence of Markov times defined as follows:

$$\eta_0 = 0, \quad \eta_{i+1} = \inf\{s > \eta_i : \nu_s \neq \nu_{\eta_i}\}; \quad n(T) = \max\{i : \eta_i \leq T\}.$$

If $n(T) = 0$ we put

$$\frac{dP_{x,k}}{d\widehat{P}_{x,k}}(X., \nu.) = \exp \left\{ -\int_0^T (\hat{c}_{\nu_0}(X_s) - n + 1) \, ds \right\}.$$

Equations (1.17) and (1.18) allow us to write

$$u_k(t,x) =$$

$$= \widehat{E}_{x,k} g_{\nu_t}(X_t) \exp\left\{\int_0^t c_{\nu_s}(t-s, X_s)\, ds\right\} \prod_{i=0}^{n(T)-1} c_{\eta_i \eta_{i+1}}(X_s) \qquad (1.19)$$

$$\exp\left\{-\sum_{i=0}^{n(T)} \int_{\eta_i}^{\eta_{i+1}\wedge T} \left[\hat{c}_{\nu_{\eta_i}}(X_s) - n + 1\right] ds\right\}.$$

Here $\widehat{E}_{x,k}$ is the expectation with respect to the measure $\widehat{P}_{x,k}$. It turns out that representation (1.19) for the solution of problem (1.15) holds without the assumption concerning nonnegativity of $c_{ij}(x)$. This can be proved easily if one notes that the right side of (1.19) defines a semigroup $(T_t g)(x, k)$ in the space of bounded measurable functions on $\mathrm{R}^r \times \{1, \ldots, n\}$, and the generator of this semigroup for smooth enough functions coincides with the right side of system (1.15) for $c_k(x) \equiv 0$, $k \in \{1, \ldots, n\}$, $x \in \mathrm{R}^r$.

It is worth noting that one can write down a system similar to (1.10) for the solutions of problem (1.13) with coefficients a_k^{ij}, b_k^i depending not just on $x \in \mathrm{R}^r$, but on the unknown functions $u_1, \ldots, u_n$, as well. Using the methods developed for a single quasilinear equation (see [F6, Ch. 5]), one can study existence, uniqueness and smoothness of the solutions of such systems and their qualitative properties. We will not study these problems here.

Consider now the Dirichlet problem:

$$L_k u_k(x) + \sum_{j=1}^{n} c_{kj}(x)(u_j - u_k) = 0, \quad x \in G \subset \mathrm{R}^r, \qquad (1.20)$$

$$u_k(x)\Big|_{\partial G} = \psi_k(x), \quad k = 1, \ldots, n.$$

Let, for brevity, $c_{kj}(x) \geq 0$, $L_k, k \in \{1, \ldots, n\}$, be non-degenerative elliptic operators. We assume that all the coefficients are Lipschitz continuous, the domain G is bounded and has a smooth boundary, and the functions $\psi_k(x)$, $k \in \{1, \ldots, n\}$, $x \in \partial G$, are continuous. If $(X_t, \nu_t; P_{x,k})$ is the Markov process defined by (1.16), and $\tau = \min\{t : X_t \notin G\}$, then the solution of (1.20) can be written in the form:

$$u_k(x) = E_{x,k} \psi_{\nu_\tau}(X_\tau), \quad x \in G, \; k \in \{1, \ldots, n\}. \qquad (1.21)$$

The proof of (1.21) can be found in [EF1].

In the case of a single equation, the representation of the solution of the linear problem (1.6) in the form of a functional integral gives an integral equation for the solution of one reaction-diffusion equation. Similarly, the representations (1.17),

(1.19) give an integral equation for the solution of the RDE system. We will write down and use such integral equations later.

Until now we considered PDE's with the Dirichlet conditions on the boundary. In the conclusion of this section we introduce the process with reflection on the boundary, which allows us to give a probabilistic representation for the solutions of problems with Neumann type boundary conditions.

Let G be a domain in $\mathbf{R}^r$ with C^3-class boundary ∂G, and $\ell(x)$ be a C^2-class vector field on ∂G not tangent to ∂G. Consider the initial boundary problem

$$\frac{\partial u(t,x)}{\partial t} = Lu + c(t,x)u, \ \ t > 0, \ x \in G$$

$$u(0,x) = g(x), \ \ \frac{\partial u(t,x)}{\partial \ell(x)}\bigg|_{\substack{x \in \partial G, \\ t > 0}} = 0. \tag{1.22}$$

Here L is a non-degenerate operator with Lipschitz continuous coefficients. To construct the process (X_t, P_x), $x \in G \cup \partial G$, governed by the operator L inside G with reflection on the boundary in the direction $\ell(x)$, consider the stochastic differential equation

$$dX_t^x = \sigma(X_t^x)\, dW_t + b(X_t^x)\, dt + \chi_{\partial G}(X_t^x)\, \ell(X_t^x)\, d\xi_t^x,$$

$$X_0^x = x \in G \cup \partial G, \ \ \xi_0^x = 0, \tag{1.23}$$

where $\chi_{\partial G}$ is the indicator function of the boundary ∂G, $\sigma(x)$ and $b(x)$ were introduced above. By a solution of (1.23) we mean a pair of a.s. continuous processes X_t^x, ξ_t^x satisfying (1.23) (including the assumption that X_t^x and ξ_t^x are measurable with respect to the σ-field generated by the variables $W_s, s \leq t$), and satisfying a.s. the following conditions:

$X_t^x \in G \cup \partial G$, ξ_t^x is a non-decreasing process which increases only at points $t \in \{s : X_s^x \in \partial G\} = \Lambda$; Λ has Lebesgue measure 0 a.s.

One can prove that under the conditions mentioned above there exists a unique solution $(X_t^x, \ \xi_t^x)$ of equation (1.23). The process (X_t, P_x) corresponding to the family (X_t^x, P) is called the process with reflection in the direction $\ell(x)$ governed by the operator L inside G. The process ξ_t^x is called local time of (X_t^x, P) on ∂G (see, for example, [F6], §1.6).

One can prove ([F6], §2.5) that the solution of the problem (1.22) can be written as follows

$$u(t,x) = E_x\, g(X_t) \exp\left\{\int_0^t c(t-s, X_s)\, ds\right\}. \tag{1.24}$$

The solution of the Neumann problem

$$Lu(x) - c(x)u(x) = f(x), \ \ x \in G, \ \ \frac{\partial u(x)}{\partial \ell(x)}\bigg|_{\partial G} = 0, \tag{1.25}$$

with $c(x) > c_0 = 0$ has the representation

$$u(x) = -\int_0^\infty E_x\, f(X_t)\, \exp\left\{-\int_0^t c(X_s)\, ds\right\} dt. \qquad (1.26)$$

If $c(x) \equiv 0$, the problem (1.25) is solvable only for $f(x)$ orthogonal to the invariant measure of the process (X_t, P_x) in $G \cup \partial G$ with reflection along field $\ell(x)$ on the boundary. For such $f(x)$ the integral in (1.26) is finite for $c(x) \equiv 0$, and formula (1.26) gives the solution of problem (1.25) ([F6], § 2.5).

One can write down similar representations for solutions of the system (1.15) or for the corresponding stationary problem for a system with Neumann type boundary conditions. For example, if we consider equations (1.15) in a bounded domain $G \subset \mathrm{R}^r$ having a smooth boundary ∂G with conditions

$$\left.\frac{\partial u_k(t, x)}{\partial l(x)}\right|_{\substack{x \in \partial G, \\ t > 0}} = 0, \;\; u_k(0, x) = g_k(x), \;\; k = 1, \ldots, n, \qquad (1.27)$$

then the corresponding process $(X_t, \nu_t, P_{x,k}) = Z$ can be defined by equations of (1.23) type supplemented with the law of evolution of ν_t. The solution of this Neumann type initial-boundary problem for systems can be represented through the process Z by formula (1.17).

2   Small Parameter in Higher Derivatives: Levinson's Case

Let

$$L^\epsilon = \frac{\epsilon^2}{2} \sum_{i,j=1}^{r} a^{ij}(x) \frac{\partial^2}{\partial x^i \partial x^j} + \sum_{i=1}^{r} b^i(x) \frac{\partial}{\partial x^i}, \quad x \in \mathbb{R}^r,$$

where the coefficients $a^{ij}(x), b^i(x)$ are assumed to be bounded and Lipschitz continuous, and $\sum_{i,j=1}^{r} a^{ij}(x)\lambda_i\lambda_j \geq a \sum_1^r \lambda_i^2$ for any real $\lambda_1, \ldots, \lambda_r$ and some $a > 0$.
Consider the process corresponding to L^ϵ:

$$\dot{X}_t^{x,\epsilon} = \epsilon\sigma(X_t^{x,\epsilon})\dot{W}_t + b(X_t^{x,\epsilon}), \quad X_0^{x,\epsilon} = x. \tag{2.1}$$

Equation (2.1) turns into an ordinary differential equation if $\epsilon = 0$:

$$\dot{X}_t^x = b(X_t^x), \quad X_0^x = x. \tag{2.2}$$

Using the Lipschitz continuity of the coefficients and Gronwall's inequality, one can easily derive that

$$\sup_{0 \leq t \leq T} |X_t^{x,\epsilon} - X_t^x| \leq e^{KT} \cdot \epsilon \sup_{0 \leq t \leq T} \left| \int_0^t \sigma(X_s^{x,\epsilon}) \, dW_s \right|, \tag{2.3}$$

where K is the Lipschitz constant. It follows from the properties of the Itô integral that the supremum on the right-hand side of (2.3) is finite a.s., and for any $a > 0$

$$P\{ \sup_{0 \leq t \leq T} | \int_0^t \sigma(X_s^\epsilon) dW_s | > a\} \leq \frac{1}{a^2} \int_0^T E(\sum_{i=1}^{r} a^{ii}(X_s^{\epsilon,x})) ds.$$

Thus, the left-hand side of (2.3) tends to zero as $\epsilon \to 0$ with probability 1 starting at any point $x \in \mathbb{R}^r$, and for any $T, \delta > 0$ there exists $N = N(T, \delta)$ such that

$$P\{ \sup_{0 \leq t \leq T} | X_t^{\epsilon,x} - X_t^x | > \delta\} \leq \epsilon^2 N(T, \delta).$$

This means that one can look on the process $X_t^{\epsilon,x}$, defined by (2.1), as a result of small random perturbations of dynamical system (2.2).

Consider now the Cauchy problem

$$\frac{\partial u^\epsilon(t,x)}{\partial t} = L^\epsilon u^\epsilon, \quad t > 0, \quad x \in \mathbb{R}^r, \quad u^\epsilon(0,x) = g(x).$$

As follows from (1.6),

$$u^\epsilon(t,x) = E_x g(X_t^\epsilon). \tag{2.4}$$

If $g(x)$ is continuous and bounded, one can conclude that

$$\lim_{\epsilon \to 0} u^\epsilon(t, x) = g(X_t^x) = u^0(t, x).$$

The function $u^0(t, x)$ is the solution (maybe generalized) of the degenerate problem

$$\frac{\partial u^0(t, x)}{\partial t} = L^0 u^0 = \sum_{i=1}^r b^i(x) \frac{\partial u^0}{\partial x^i}, \quad u^0(0, x) = g(x).$$

Assume that $g(x), b(x), \sigma(x)$ are infinitely differentiable and bounded together with their derivatives. Then one can prove (see [BF], [FW1]), that

$$X_t^{x,\epsilon} = X_t^x + \epsilon X_t^{x,(1)} + \cdots + \epsilon^n X_t^{x,(n)} + \cdots. \tag{2.5}$$

One can derive from (2.4) and (2.5) the expansion for $u^\epsilon(t, x)$:

$$u^\epsilon(t, x) = u^0(t, x) + \epsilon^2 u^{(1)}(t, x) + \cdots + \epsilon^{2k} u^{(k)}(t, x) + \cdots.$$

The odd powers of ϵ appear in the case of finite smoothness of the data. For example, if g is only Lipschitz continuous then the difference $u^\epsilon(t, x) - u^0(t, x)$ will be of order ϵ as $\epsilon \downarrow 0$.

A much more interesting situation arises in the case of the Dirichlet problem for an equation with a small parameter in higher derivatives:

$$L^\epsilon u^\epsilon(x) = \frac{\epsilon^2}{2} \sum_{i,j=1}^r a^{ij}(x) \frac{\partial^2 u^\epsilon}{\partial x^i \, \partial x^j} + \sum_{i=1}^r b^i(x) \frac{\partial u^\epsilon}{\partial x^i} = 0, \quad x \in G,$$

$$u^\epsilon(x)\Big|_{\partial G} = \psi(x). \tag{2.6}$$

We assume for brevity that $G \subset \mathbf{R}^r$ is a bounded domain with a smooth boundary ∂G and that the boundary function $\psi(x)$ is also smooth. The solution of problem (2.6) has the representation

$$u^\epsilon(x) = E_x \psi(X_{\tau^\epsilon}^\epsilon), \quad \tau^\epsilon = \min\{t : X_t^\epsilon \in \partial G\}. \tag{2.7}$$

To understand the behavior of $u^\epsilon(x)$ as $\epsilon \downarrow 0$, one should study $X_{\tau^\epsilon}^\epsilon$ as $\epsilon \downarrow 0$. Various cases are possible depending on the behavior of the trajectories of dynamical system (2.2) starting from $x \in G$. First, it can happen that X_t^x, $x \in G$, leaves G in a finite time: $T(x) = \min\{t : X_t^x \notin G\} < \infty$ for $x \in G$ (Fig. 1).

One says that Levinson's condition is fulfilled if $T(x) < \infty$ for $x \in G$, and X_t^x, $x \in G$, leaves the domain G immediately after $T(x)$: for each $x \in G$ there exists $\delta > 0$ such that $X_{T(x)+s}^x \notin G \cup \partial G$ for $s \in (0, \delta]$.

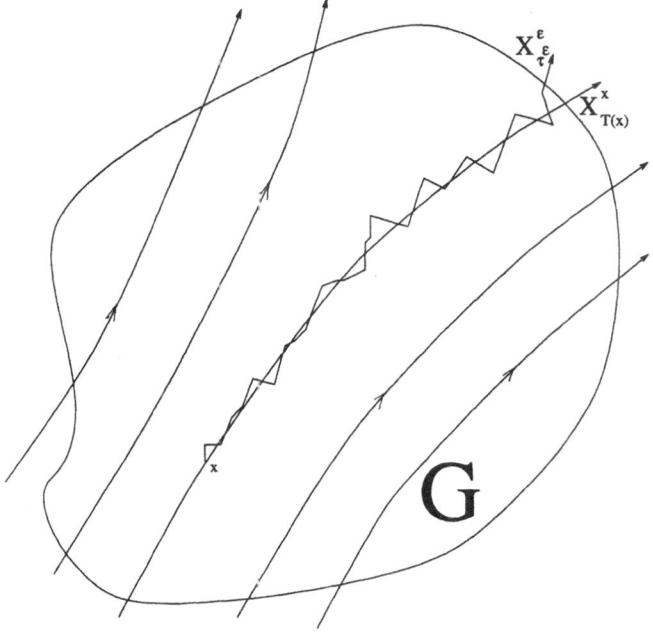

Figure 1

It is easy to check that if the Levinson condition is fulfilled, then $u^G(x) = \psi(X^x_{T(x)})$ is the unique (maybe generalized) solution of the problem

$$\sum_{i=1}^{r} b^i(x)\frac{\partial u^0(x)}{\partial x^i} = 0, \ x \in G, \ u^0(x)\Big|_{\widetilde{\partial G}} = \psi(x), \tag{2.8}$$

where $\widetilde{\partial G}$ is the part of the boundary which is regular for L^0: $\widetilde{\partial G} = \{y \in \partial G : X^y_t \notin G \cup \partial G$ for small enough $t > 0\}$.

It was explained above that for any $T > 0$

$$P_x\left\{\max_{0 \le t \le T} | X^{\epsilon,x}_t - X^x_t | > \delta\right\} \to 0 \ \text{as} \ \epsilon \to 0. \tag{2.9}$$

The Levinson condition and (2.9) imply that

$$| X^{\epsilon,x}_{\tau^\epsilon} - X^x_{T(x)} | \to 0$$

in probability as $\epsilon \downarrow 0$. This relation and the continuity of the boundary function $\psi(x)$ result in the convergence

$$u^\epsilon(x) = E_x\psi(X^\epsilon_{\tau^\epsilon}) \to \psi(X^x_{T(x)}) = u^0(x), \ \epsilon \downarrow 0.$$

This result, using analytical methods, was first proved by N. Levinson in 1950.

If the coefficients of the equation, the boundary and the boundary functions are infinitely differentiable, one can prove that an asymptotic expansion exists in the Levinson case:

$$u^\epsilon(x) = u^0(x) + \epsilon^2 u^{(1)}(x) + \epsilon^4 u^{(2)}(x) + \dots.$$

In particular, one can check that

$$\mid u^\epsilon(x) - u^0(x) \mid < \epsilon^2 \cdot \text{const.}$$

in any compact subdomain of G.

We will come back to the Levinson case later in a more complicated situation. Now I want to mention the other cases. The exit of $X_t^{\epsilon,x}$ from G in the Levinson case occurs due to dynamical system (2.2). We face another generic case when the dynamical system hinders exit from the domain (Fig. 2).

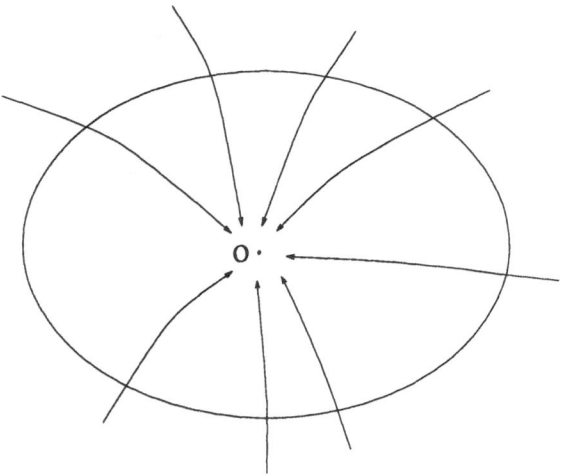

Figure 2

If the vector field $b(x)$ is directed strictly inside G, the exit of $X_t^{\epsilon,x}$ from G will also occur with probability 1 for any $x \in G$ and $\epsilon > 0$, but it will take much more time. The trajectory $X_t^{\epsilon,x}$ will again and again return to a small neighborhood of the point 0 if $\epsilon \ll 1$. The exit time τ^ϵ will be logarithmically equivalent to $\exp\{\frac{c}{\epsilon^2}\}$ for some $c > 0$ as $\epsilon \downarrow 0$. The exit of $X_t^{\epsilon,x}$ from G in this case is due to the perturbations. More precisely, it occurs due to large deviations of the process from its typical behavior.

Finally, one more case should be mentioned. Assume that the dynamical system behaves as shown in Fig. 3:

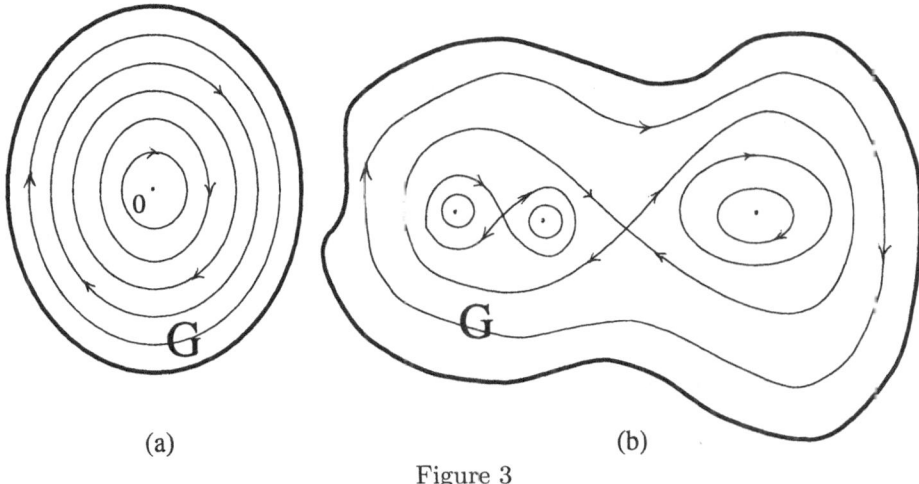

(a)   (b)

Figure 3

The vector field $b(x)$ does not help to exit the domain, but does not hinder the exit either. Suppose the domain $G \subset \mathrm{R}^2$ in Fig. 3 is bounded by a trajectory of system (2.2), and the trajectories behave like in Fig. 3a or 3b inside G. Then the process (X_t^ϵ, P_x) for $\epsilon \ll 1$ will make many rotations along the periodic trajectories of the dynamical system before it moves a little in the direction transversal to the periodic trajectories. The characteristics of the motion in the transversal direction (slow motion) should be averaged in the fast motion along the periodic non-random trajectories. In the case of Fig. 3a these arguments allow us to give a solution of the problem: to describe the limiting slow motion in an appropriate time scale [Xh1]. In the case of Fig. 3b, as we will see later, these arguments are not sufficient. To preserve the Markov property for the limiting slow motion, one should consider the slow motion on a graph corresponding to the dynamical system shown in Fig. 3b and describe the process at the vertices (See section 5).

The three cases mentioned above – Levinson's case, large deviations case, and the case of averaging – are extreme cases. Of course, a mixture of them is also possible.

Consider now the Dirichlet problem for a system of type (1.20) with a small parameter in the higher derivatives:

$$L_k^\epsilon u_k^\epsilon + \sum c_{kj}(x)(u_j^\epsilon - u_k^\epsilon) = 0, \quad x \in G \subset \mathrm{R}^r,$$

$$u_k^\epsilon(x) \mid_{\partial G} = \psi_k(x), \quad k = 1, \ldots, n. \tag{2.10}$$

Here

$$L_k^\epsilon = \frac{\epsilon^2}{2} \sum_{i,j=1}^r a_k^{ij}(x) \frac{\partial^2}{\partial x^i \partial x^j} + \sum_{i=1}^r b_k^i(x) \frac{\partial}{\partial x^i}, \quad k = 1, \ldots, n.$$

It turns out that even in the simplest case, when a counterpart of the Levinson condition is fulfilled, the situation becomes more complicated. More precisely, there are two natural ways to generalize the Levinson condition. In the case of one equation the process (X_t^ϵ, P_x) for $\epsilon = 0$ turns into the deterministic dynamical system (2.2). The exit time $T(x)$ for this system is non-random. The process $(X_t^\epsilon, \nu_t^\epsilon; P_{x,k})$ turns into the random process $(X_t^0, \nu_t^0; P_{x,k})$, corresponding to the first order system

$$\sum_{i=1}^{r} b_k^i(x) \frac{\partial u_k^0}{\partial x^i} + \sum_{j=1}^{r} c_{kj}(x)(u_j^0 - u_k^0) = 0, \ x \in \mathrm{R}^r, \ k = 1, \ldots, n.$$

This process consists of deterministic motion along one of the fields $b_k(x) = (b_k^1(x), \ldots, b_k^r(x))$ and of random switchings from one field to another. Denote

$$\tau^0(x, k) = \min\{t : X_t^{0,x,k} \in \partial G\}.$$

Now $\tau^0(x, k)$, the exit time, starting from $x \in G$, $k \in \{1, \ldots, n\}$ is a random variable.

We say that a weak Levinson's condition is fulfilled, if:

(i) there exist $T_0 > 0$ and $\delta_0 > 0$ such that $P_{x,k}(\tau^0 < T_0) > \delta_0$ for any $x \in G$, $k = 1, \ldots, n$;

(ii) for any $k \in \{1, \ldots, n\}$ there exist sets $\Gamma_k^+, \Gamma_k^- \subset \partial G$ such that

$$\partial G = \Gamma_k^+ \cup \Gamma_k^-, (b_k(x), n(x)) \leq -\alpha_0 < 0 \text{ for } x \in \Gamma_k^-$$

$$(b_k(x), n(x)) \geq \alpha_0 > 0 \text{ for } x \in \Gamma_k^+,$$

where α_0 is independent of $x \in \partial G$ and k, and $n(x)$ is the outward normal at $x \in \partial G$.

Using the Markov property one can derive from the first condition that T_1, $c_0 > 0$ exist such that for any $x \in G$, $k \in \{1, \ldots, n\}$ and $t > T_1$

$$P_{x,k}\{\tau^0 > t\} \leq \exp\{-tc_0\}, \tag{2.11}$$

and, of course, $P_{x,k}\{\tau^0 < \infty\} = 1$, $x \in G$, $k \in \{1, \ldots, n\}$. Thus the weak Levinson condition causes the process to exit from the domain within a finite time. Note that the best possible constant c_0 for which the bound (2.11) still holds is the first eigenvalue of the problem:

$$L_k^0 \psi_k(x) + \sum_{j=1}^{n} c_{kj}(\psi_j - \psi_k) = \lambda \psi_k(x), \ x \in G, \ \psi_k(x) \mid_{\Gamma_k^+} = 0, \ k = 1, \ldots, n. \tag{2.12}$$

The second condition is rather restrictive, and it can be replaced by a less restrictive one. But our goal here is to demonstrate some effects in the simplest situation rather than to consider the most general case. Thus we assume conditions (i) and (ii) when we say that the Levinson conditions are fulfilled. Moreover, let us assume here for brevity that the coefficients c_{ij} are positive constants. The general case can be reduced to this one using absolute continuity of measures corresponding to processes with different coefficients $c_{ij}(x)$ (see [EF1]).

The solution of problem (2.10) can be written as follows

$$u_k^\epsilon(x) = E_{x,k}\psi_{\nu_{\tau^\epsilon}}(X_{\tau^\epsilon}^\epsilon), \tag{2.13}$$

where

$$\dot{X}_t^{\epsilon,x,k} = \epsilon\sigma_{\nu_t^k}(X_t^{\epsilon,x,k})\dot{W}_t + b_{\nu_t^k}(X_t^{\epsilon,x,k}), \quad X_0^{\epsilon,x,k} = x, \tag{2.14}$$

ν_t^k is the continuous time Markov chain with phase space $\{1,\ldots,n\}$ and transition intensities $c_{ij} > 0$, starting at $\nu_0^k = k$, $\tau^\epsilon = \tau^\epsilon(x,k) = \min\{t : X_t^{\epsilon,x,k} \in \partial G\}$.

It is easy to check [EF1] that under the Levinson condition the degenerate problem

$$L^0 u_k^0(x) = \sum_{i=1}^{r} b_k^i(x)\frac{\partial u_k^0}{\partial x^i} + \sum_{j=1}^{n} c_{kj} \cdot (u_j^0 - u_k^0) = 0, \quad x \in G,$$

$$u_k^0(x)\Big|_{x\in\Gamma_k^+} = \psi_k(x), \quad k \in \{1,\ldots,n\},$$

has unique solution (maybe generalized). This solution has the representation:

$$u_k^0(x) = E_{x,k}\psi_{\nu_{\tau^0}}(X_{\tau^0}^0),$$

where X_t^0 is defined by

$$\dot{X}_t^{0,x,k} = b_{\nu_t^k}(X_t^{0,x,k}), \quad X_0^{0,x,k} = x; \quad \tau^0(x,k) = \min\{t : X_t^{0,x,k} \in \partial G\}. \tag{2.15}$$

Theorem 2.1 *Assume that the Levinson condition is fulfilled and $\psi_k(x)$, $1 \leq k \leq n$, $x \in \partial G$, are Hölder continuous. Then for any $\gamma_2 \in (0,1/2)$ there exist $\gamma_1 > 0$ and $\epsilon_0 > 0$ such that for $\epsilon \leq \epsilon_0$ and $x \in G$ with $\mathrm{dist}(x,\partial G) \geq \epsilon^{\gamma_2}$ the following bound holds:*

$$\mid u_k^\epsilon(x) - u_k^0(x) \mid \leq \epsilon^{\gamma_1}. \tag{2.16}$$

Let us first consider an example that shows that the bound (2.16) cannot be improved in the class of all systems satisfying weak Levinson's conditions, and that the difference in the left side of (2.16) can be bigger than ϵ^γ for arbitrary small $\gamma > 0$ even if the data of the problem are infinitely smooth.

Example. Let $D = (-1, 2) \subset R^1$, and consider the problem

$$\frac{\epsilon^2}{2} \frac{d^2 u_1^\epsilon}{dx^2} + b_1(x) \frac{du_1^\epsilon}{dx} + u_2^\epsilon - u_1^\epsilon = 0,$$

$$\frac{\epsilon^2}{2} \frac{d^2 u_2^\epsilon}{dx^2} + b_2(x) \frac{du_2^\epsilon}{dx} + u_1^\epsilon - u_2^\epsilon = 0, \qquad (2.17)$$

$$x \in D, \quad u_1^\epsilon(-1) = u_2^\epsilon(-1) = 0,$$

$$u_1^\epsilon(2) = u_2^\epsilon(2) = 1.$$

Put $b_1(x) = \alpha x$, $b_2(x) = \alpha \cdot (x - 1)$, where α is a positive constant.

Each of the functions $b_i(x)$ is equal to zero at some point of the domain D. It is easy to check however that the Levinson conditions are satisfied here due to the switchings of the component ν_t.

Now, since $b_i(x) \leq 0$ for $x \in [-1, 0]$ and $i = 1, 2$,

$$P_{x,1}\{X^0(\tau^0) = 2\} = 0 \text{ for } x \leq 0.$$

Here $\tau^0 = \min\{t : X_t^0 \notin (-1, 2)\}$. Thus

$$u_1^0(x) = P_{x,1}\{X^0(\tau^0) = 2\} = 0 \text{ for } x \leq 0. \qquad (2.18)$$

On the other hand, for $\epsilon > 0$,

$$u_1^\epsilon(0) = P_{0,1}\{X^\epsilon(\tau^\epsilon) = 2\} \geq$$

$$\geq P_{0,1}\left\{X^\epsilon(\tau^\epsilon) = 2, \; \eta_1 \geq \frac{4}{\alpha}\ln\frac{1}{\epsilon} + 1, \; \tau^\epsilon \leq 1 + \frac{4}{\alpha}\ln\frac{1}{\epsilon}\right\}, \qquad (2.19)$$

where $\eta_1 = \min\{t : \nu_t \neq 1\}$ is the time of the first jump of the component ν_t. It is clear that for $t \in [0, \eta_1]$ the trajectory $X_t^{\epsilon;0,1}$ coincides with Y_t^ϵ:

$$Y_t^{\epsilon;0} = \alpha Y_t^{\epsilon;0} + \epsilon \dot{W}_t, \; Y_0^{\epsilon;0} = 0. \qquad (2.20)$$

Therefore, taking into account independence of η_1 and $Y_\cdot^\epsilon$, we conclude from (2.18):

$$u_1^\epsilon(0) \geq P_0\left\{Y_{\tau_1^\epsilon}^\epsilon = 2, \; \tau_1^\epsilon < \frac{4}{\alpha}\ln\frac{1}{\epsilon} + 1\right\} \cdot P\left\{\eta_1 \geq 1 + \frac{4}{\alpha}\ln\frac{1}{\epsilon}\right\}. \qquad (2.21)$$

The random variable η_1 has exponential distribution with mean value 1. Thus the second factor on the right hand side of (2.20) is equal to

$$\exp\left\{-(1 + \frac{4}{\alpha}\ln\frac{1}{\epsilon})\right\} = e^{-1}\epsilon^{\frac{4}{\alpha}}.$$

Process $Y_t^{\epsilon,0}$ defined by (2.19) has a very simple structure, and it is not difficult to check that the first factor in the right side of (2.20) is bounded from below by a constant $A > 0$ independent of ϵ for ϵ small enough (see [F6], §4.2). We conclude from (2.20) that $u_1^\epsilon(0) > Ae^{-1} \cdot \epsilon^{\frac{4}{\alpha}}$. This bound together with (2.17) implies:

$$| u_1^\epsilon(0) - u_1^0(0) | > A_1 \epsilon^{4/\alpha}.$$

Thus the convergence of $u^\epsilon(x)$ to the limiting function $u^0(x)$ in the case of systems, even if the Levinson conditions are fulfilled, can be very slow, and bound (2.16) cannot be improved in the whole class of system satisfying the Levinson conditions.

In our example the fields $b_k(x)$ have equilibrium points inside D. There are examples where each of the fields $b_k(x)$ satisfies the Levinson condition for one equation, but the difference $u_k^\epsilon(x) - u_k^0(0)$ can be bounded from below by ϵ^γ for arbitrary small γ [EF1].

Let us now outline the proof of Theorem 2.1 [EF1]. Consider, for brevity, only convergence in $\widetilde{G} \subset G$ such that $\text{dist}(\widetilde{G}, \partial G)$ is positive and independent of ϵ. We use the representations

$$u_k^\epsilon(x) = E_{x,k}\psi_{\nu_{\tau\epsilon}}(X_{\tau^\epsilon}^\epsilon), \quad u_k^0(x) = E_{x,k}\psi_{\nu_{\tau^0}}(X_{\tau^0}^0). \tag{2.22}$$

Here ν_t is, independent of the ϵ Markov chain, the same in both expectations. The trajectories $X^{\epsilon;x,k}$ and $X^{0;x,k}$ are defined by equations (2.14) and (2.15). Subtracting (2.15) from (2.14), we have:

$$X_t^{\epsilon;x,k} - X_t^{0;x,k} = \int_0^t \left[b_{\nu_s^k}(X_s^{\epsilon;s,k}) - b_{\nu_s^k}(X_s^{0;x,k})\right] ds +$$
$$+\epsilon \int_0^t \sigma_{\nu_s^k}(X_s^{\epsilon;x,k})dW_s. \tag{2.23}$$

Since the fields $b_k(x)$ are Lipschitz continuous with a constant K, we derive from (2.23):

$$\max_{0 \le s \le t} \left|X_s^{\epsilon;x,k} - X_s^{0;x,k}\right| \le e^{Kt}\epsilon \max_{0 \le t_1 \le t} \left|\int_0^{t_1} \sigma_{\nu_s^k}(X_s^{\epsilon;x,k})dW_s\right|. \tag{2.24}$$

Using standard properties of Itô's integral, one can obtain from (2.23):

$$E\left[\max_{0 \le s \le t} \left|X_s^{\epsilon;x,k} - X_s^{0;x,k}\right|^2\right] \le \epsilon^2 \cdot e^{2Kt}A_1 t, \tag{2.25}$$

where A_1 is a constant depending on the $\max_{\substack{x \in G \cup \partial G \\ i,j,k}} \left|a_k^{ij}(x)\right|$.

Consider now the exit times from G, $\tau^{\epsilon;x,k}$ and $\tau^{0;x,k}$, for trajectories $X_t^{\epsilon;x,k}$ and $X_t^{0;x,k}$ and

$$\bar{\tau} = \tau^{\epsilon;x,k} \wedge \tau^{0;x,k}.$$

The variable $\bar{\tau}$, as well as $\tau^{\epsilon;x,k}$ and $\tau^{0,x,k}$, are Markov times. Thus, we can use the strong Markov property and conclude from (2.21):

$$u_k^\epsilon(x) = Eu_{\nu_{\bar{\tau}}^k}^\epsilon(X_{\bar{\tau}}^{\epsilon;x,k}), \quad u_k^0(x) = Eu_{\nu_{\bar{\tau}}^k}^0(X_{\bar{\tau}}^{0;x,k}). \tag{2.26}$$

Denoting by χ_A the indicator function of a set A, we have from (2.25):

$$\left|u_k^\epsilon(x) - u_k^0(x)\right| \le \left|E\left[\psi_{\nu_{\bar{\tau}}^k}\left(X_{\bar{\tau}}^{\epsilon,x,k}\right) - u_{\nu_{\bar{\tau}}^k}^0\left(X_{\bar{\tau}}^{0,x,k}\right)\right]\chi_{\bar{\tau}=\tau^\epsilon}\chi_{\bar{\tau}\le T}\right|$$

$$+ \left|E\left[u_{\nu_{\bar{\tau}}^k}^\epsilon\left(X_{\bar{\tau}}^{\epsilon,x,k}\right) - \psi_{\nu_{\bar{\tau}}^k}\left(X_{\bar{\tau}}^{0,x,k}\right)\right]\chi_{\bar{\tau}=\tau^0}\chi_{\bar{\tau}\le T}\right| + \tag{2.27}$$

$$+ A_2 P\{\bar{\tau} \ge T\},$$

where A_2 is a constant defined by $\max_{k,x}|\psi_k(x)|$.

Now, let $T = \alpha \ln \frac{1}{\epsilon}$; the constant $\alpha > 0$ will be chosen later. Then, because of (2.24),

$$E\left|X_{\bar{\tau}\wedge T}^{\epsilon,x,k} - X_{\bar{\tau}\wedge T}^{0,x,k}\right|^2 < A_3\epsilon^{2-2\alpha K}\ln \epsilon^{-1}. \tag{2.28}$$

One can derive from assumption (ii) of the Levinson condition, that

$$P\left\{X_{\tau^\epsilon}^{\epsilon,x,k} \in \bigcup_k \Gamma_k^-\right\} < A_4\epsilon, \ \epsilon \downarrow 0, \ x \in \tilde{G}. \tag{2.29}$$

If $z \in \Gamma^+$ and $x \in G$, then taking into account (ii) and Hölder continuity of the boundary functions, we obtain:

$$|\psi_k(z) - u_k^\epsilon(x)| \le E\left|\psi_k(z) - \psi_k\left(X_{\tau^\epsilon}^{\epsilon,x,k}\right)\right| +$$

$$+ \max_{x,i}|\psi_i(x)|P\left\{\nu_t^k \text{ had a jump between 0 and } \tau^{\epsilon,x,k}\right\} \le$$

$$\le A_5 E\left|z - X_{\tau^\epsilon}^{\epsilon,x,k}\right|^\mu + A_6|z - x| \le \tag{2.30}$$

$$\le A_5\left[|z - x|^2 + |x - X_{\tau^\epsilon}^{\epsilon,x,k}|^2\right]^{\mu/2} + A_6|z - x| \le$$

$$\le A_7|z - x|^{\mu/2},$$

for $|z - x|$ small enough; $\mu \le 1$ here is the Hölder exponent of the boundary function.

Combining (2.26)–(2.29), (2.11) and choosing $\alpha = \frac{\mu}{c_0 + K\mu}$, we obtain:

$$|u_k^\epsilon(x) - u_k^0(x)| \le A_8(\epsilon^{\mu-\alpha K\mu} + \epsilon^{c_0\alpha})\ln \epsilon^{-1} \le$$

$$\le A_9\epsilon^\gamma \text{ for } \gamma < \frac{\mu c_0}{K\mu + c_0}.$$

Thus, one can guarantee the rate of convergence of u_k^ϵ to u_k^0 of order ϵ^γ for some $\gamma > 0$. Even if the boundary function is infinitely differentiable the rate of convergence can be small. This rate is determined by a competition between the rate of divergence of trajectories $X_t^{\epsilon,x,k}$ and $X_t^{0,x,k}$ defined by the Lipschitz constant K [see (2.23), (2.24)] and the rate of decreasing of the probabilities $P_{x,k}\{\tau^\epsilon > t\}$ as $t \to \infty$ given by (2.11).

The interplay between these two exponential bounds, by the way, is essential in many problems. For example, smoothness of solutions of degenerate equations and existence of solutions of quasilinear degenerate equations which are continuous for all $t > 0$ are determined by this interplay. See [F6], Chapters 3, 5, for the case of a single degenerate equation. Similar results can be proved for systems.

A similar effect of slow convergence of the solution of a Dirichlet problem with a small parameter to the solution of the degenerate problem under the Levinson condition can be observed also in the case of one equation. It happens if not all second derivatives in the equation have the small factor. A bound for the rate of convergence and some examples for such equations one can be found in [F6] § 4.1.

One should say, as the examples show, that our estimates (and the estimates for one equation in [F6] mentioned above) are precise if we do not make special assumptions on the structure of the degenerate equation (other than the Levinson conditions). One can prove that if the coefficients and the boundary functions are smooth enough, then the rate of convergence will be of order ϵ^2, as in the classical case, provided that the ratio c_0/K is sufficiently large. Moreover, if c_0/K is large enough, then the next terms of the asymptotic expansion for $u_k^\epsilon(x) - u_k^0(x)$ can be calculated. On the other hand, it is easy to prove that, if we consider perturbations of a non-degenerate second order operator, then the rate of convergence will be of order ϵ^2 without any assumptions concerning the ratio $\frac{c_0}{K}$. An interesting open question is: what kind of conditions on the non-perturbed operator provide the rate of convergence of order ϵ^2 even if $\frac{c_0}{K}$ is small. This question is, of course, closely connected with the local smoothness of the solutions. One can expect that it should be the Hörmander type conditions. The problem is interesting even in the case of a single equation. Another open question is the following. Find a function $\alpha(\epsilon)$ such that there exist a finite nonzero limit $\lim_{\epsilon \downarrow 0}(u_k^\epsilon(x) - u_k^0(x))\alpha(\epsilon)$.

One can introduce a strong Levinson condition, replacing condition (i) by (i′):
(i′) There exists $T > 0$ such that

$$P_{x,k}\{\tau^0 < T\} = 1 \text{ for } x \in G \cup \partial G, \ 1 \leq k \leq n.$$

For example, condition (i′) is fulfilled if all fields $b_k(x)$, $1 \leq k \leq n$, have a projection on a straight line which is separated from zero for all x and k , if and the domain G is bounded. One can prove that under a strong Levinson's condition $\mid u_k^\epsilon(x) - u_k^0(x) \mid \leq$ const·ϵ, if the boundary functions are Lipschitz continuous. One can write down the asymptotic expansion for $u_k^\epsilon(x) - u_k^0(x)$ in the even powers of the small parameter if the coefficients and boundary functions are smooth enough.

In the conclusion of this section we will mention one more small parameter problem where a counterpart of the Levinson condition arises. Consider the Dirichlet problem

$$L_k^\epsilon u^\epsilon = \frac{\epsilon^2}{2} \sum_{i,j=1}^r a_k^{ij}(x) \frac{\partial^2 u_k^\epsilon(x)}{\partial x^i \partial x^j} +$$

$$+ \epsilon \sum_{i=1}^r b_k^i(x) \frac{\partial u_k^\epsilon(x)}{\partial x^i} + \sum_{j=1}^n c_{kj}(x)(u_j^\epsilon(x) - u_k^0(x)) = 0 \quad (2.31)$$

$$u_k^\epsilon(x) \mid_{\partial G} = \psi_k(x), \; x \in G \subset R^r, \; k = 1, \ldots, n.$$

The same assumptions concerning the domain, the coefficients and the boundary functions are made as before. Of course, the solution of problem (2.31) will not change if we divide all the equations in (2.31) by ϵ. Let $(X_t^\epsilon, \nu_t^\epsilon; P_{x,k})$ be the process corresponding to the family of operators $(\frac{1}{\epsilon} L_1^\epsilon, \frac{1}{\epsilon} L_2^\epsilon, \ldots, \frac{1}{\epsilon} L_n^\epsilon)$ and $\tau^\epsilon = \min\{t : X_t^\epsilon \notin G\}$. Then the solution of (2.31) can be represented as follows:

$$u_k^\epsilon(x) = E_{x,k} \psi_{\nu_{\tau^\epsilon}^\epsilon}(X_{\tau^\epsilon}^\epsilon). \quad (2.32)$$

Using the usual notations, one can describe the component X_t^ϵ by the equation

$$\dot{X}_t^{\epsilon,x} = \epsilon^{\frac{1}{2}} \sigma_{\nu_t^\epsilon}(X_t^{\epsilon,x}) \dot{W}_t + b_{\nu_t^\epsilon}(X_t^{\epsilon,x}), \; X_0^{\epsilon,x} = x.$$

Before $X_t^{\epsilon,x}$ changes a bit, the second component makes many jumps if $\epsilon \ll 1$. Thus the distribution of ν_t^ϵ will be close to the stationary distribution $\gamma_1(X_t^{\epsilon,x}), \ldots,$ $\gamma_n(X_t^{\epsilon,x})$ of the Markov chain in the phase space $(1, \ldots, n)$ with the transition intensities $c_{ij}(X_t^{\epsilon,x})$. This implies that $X_t^{\epsilon,x}$, $0 \le t \le T$, for small ϵ is close to $\bar{X}_t^x$ in the uniform topology. Here $\bar{X}_t^x$ is the trajectory of the averaged system:

$$\dot{\bar{X}}_t^x = \bar{b}(\bar{X}_t^x), \; \bar{X}_0^x = x, \; \bar{b}(x) = \sum_k \gamma_k(x) b_k(x). \quad (2.33)$$

Similar to problem (2.6), we have the Levinson case if the trajectories $\bar{X}_t^x$ leave the domain G in a finite time $T(x)$ and cross ∂G in a regular way. In this case $X_{\tau^{\epsilon,x}}^{\epsilon,x} \to X_{T(x)}^x$ as $\epsilon \downarrow 0$ in probability. But now this not enough to calculate $\lim_{\epsilon \downarrow 0} u_k^\epsilon(x)$. As one can see from (2.31), it is necessary to know the limiting distribution of $\nu_{\tau^\epsilon}^\epsilon$ as $\epsilon \downarrow 0$. Here a number of interesting effects appear (see [FL1]).

We have the large deviation case if the field $\bar{b}(x)$ is directed strictly inside G. And, eventually, we can have the situation where the field $\bar{b}(x)$ does not hinder the exit but does not help either (see Fig. 3). We will return to this problem in the end of the next section.

3   The Large Deviation Case

The Levinson conditions mean, roughly speaking, that the trajectories of the degenerate process (or of the dynamical system, in the case of one equation) leave the domain G with probability 1, and that some regularity conditions on the boundary are fulfilled. In this case the degenerate problem ($\epsilon = 0$) with boundary conditions preserved on the regular (for the degenerate equation) part of the geometric boundary of the domain, has a unique solution. The solution of the perturbed problem in the Levinson case converges to the unique solution of the degenerate problem as $\epsilon \downarrow 0$.

Suppose now that the degenerate process never leaves the domain. More precisely, assume that the fields $b_k(x) = (b_k^1(x), \ldots, b_k^r(x))$, $k = 1, \ldots, n$, satisfy the conditions:

$$(b_k(x), n(x)) < 0, \quad k = 1, \ldots; n, \quad x \in \partial G, \tag{3.1}$$

where $n(x)$ is the outward normal to ∂G. It is easy to check that in this case the trajectories of the degenerate process $(X_t^0, \nu_t^0; P_{x,k})$ starting from $X_0^0 = x \in G$, $\nu_0^0 = k \in \{1, \ldots, n\}$ never leave $G \times \{1, \ldots, n\}$. The trajectories of the perturbed process $(X_t^\epsilon, \nu_t^\epsilon; P_{x,k})$, $\epsilon > 0$, starting from any initial point x, k, leave the domain with probability 1 and the solution of the perturbed problem is unique. In this section we study $\lim_{\epsilon \downarrow 0} u_k^\epsilon$ if conditions (3.1) are fulfilled. The limiting behavior of $u_k^\epsilon(x)$ as $\epsilon \downarrow 0$ in this case is defined by the large deviations for the corresponding family of processes.

Let us first introduce the main notions of the large deviation theory, which will be used heavily throughout these lectures.

Let X be a metric space with metric ρ, and let μ^ϵ be a family of probability measures on the Borel σ-field of X; $\epsilon > 0$ is a parameter. Let $\lambda(\epsilon)$ be a positive function, such that $\lim_{\epsilon \downarrow 0} \lambda(\epsilon) = \infty$ and let $S(x)$ be a function defined on X with values in $[0, \infty]$.

We say $S(x)$ is a normalized action function (rate function) for the family μ^ϵ in X as $\epsilon \downarrow 0$, if the following assertions hold:

(i)   for any $\delta > 0$, $\gamma > 0$ and any $x \in X$ there exists $\epsilon_0 > 0$ such that

$$\mu^\epsilon \{y : \rho(x, y) < \delta\} \geq \exp\{-\lambda(\epsilon)(S(x) + \gamma)\}$$

for $\epsilon < \epsilon_0$;

(ii)   the set $\Phi(s) = \{x \in X : S(x) \leq s\}$ is compact for every $s \geq 0$;

(iii)   for any $\delta > 0$, and $\gamma > 0$ and any $s > 0$ there exists $\epsilon_0 > 0$ such that

$$\mu^\epsilon \{y : \rho(y, \Phi(s)) \geq \delta\} \leq \exp\{-\lambda(\epsilon)(s - \gamma)\}$$

for any $\epsilon \leq \epsilon_0$.

The function $\lambda(\epsilon)$ is called the normalizing coefficient. As a rule, the measure μ^ϵ will be in our problems the distribution in a functional space X corresponding to a family of stochastic processes Z^ϵ. Then we call $S(\varphi)$, $\varphi \in X$, action functional for the family Z^ϵ as $\epsilon \downarrow 0$. One can find other equivalent definitions and the properties of action function in §3.3 of [FW1].

Consider some examples that will be used later. Let μ^ϵ be the measure in $C_{0T}(\mathrm{R}^r), T > 0$, corresponding to the family of processes $X_t^\epsilon = \epsilon W_t$, where W_t, $t > 0$, is an r-dimensional Wiener process. Then $\lambda(\epsilon) = \epsilon^{-2}$ and the normalized action functional in $C_{0T}(\mathrm{R}^r)$ as $\epsilon \downarrow 0$ is given as follows:

$$S_{0T}(\varphi) = \begin{cases} \dfrac{1}{2} \displaystyle\int_0^T |\dot\varphi_s|^2 \, ds, \ \ \varphi \in C_{0T}(\mathrm{R}^2), \ \varphi \text{ is absolutely cont.}, \ \varphi_0 = 0; \\[2mm] +\infty, \ \text{ for the rest of } C_{0T}(\mathrm{R}^r). \end{cases}$$

(3.2)

This result was first proved by M. Schilder (See also [FW1] §3.1).

For the family of diffusion processes X_t^ϵ defined by the equation

$$\dot X_t^\epsilon = \epsilon\, \sigma(X_t^\epsilon) + b(X_t^\epsilon), \ \ X_0^\epsilon = x \in \mathrm{R}^r$$

the action functional in $C_{0T}(\mathrm{R}^r)$ as $\epsilon \downarrow 0$ has the form

$$S_{0T}(\varphi) = \begin{cases} \dfrac{1}{2} \displaystyle\int_0^t \sum_{i,j=1}^r a_{ij}(\varphi_s)(\dot\varphi_s^i - b^i(\varphi_s))(\dot\varphi_s^j - b^j(\varphi_s)) \, ds, \ \text{ for} \\[2mm] \hspace{3cm} \varphi \in C_{0T}(\mathrm{R}^r), \ \varphi_0 = x, \ \varphi_s \text{ absolutely cont.}; \\[2mm] +\infty, \ \text{ for the rest of } C_{0T}(\mathrm{R}^r). \end{cases}$$

(3.3)

Here $(a_{ij}(x)) = (a^{ij}(x))^{-1}$, $(a^{ij}(x)) = \sigma(x)\sigma^*(x)$; the matrix $(a^{ij}(x))$ is supposed to be non-degenerate. The normalizing coefficient is equal to ϵ^{-2}. This was proved by S. R. S. Varadhan for $b(x) \equiv 0$ and by Freidlin and Wentzell in the general case (see [FW1], [V1]).

Consider now the Markov process $(X_t^\epsilon, \nu_t^\epsilon, P_{x,k})$ in $\mathrm{R}^r \times \{1, \ldots, n\}$, corresponding to the system

$$\frac{\epsilon^2}{2} \Delta u_k^\epsilon + \sum_{i=1}^r b_k^i(x) \frac{\partial u_k^\epsilon}{\partial x^i} + \sum_{j=1}^n c_{kj}(x)(u_j^\epsilon - u_k^\epsilon) = L^\epsilon u_k + \sum_{j=1}^n c_{kj}(x)(u_j^\epsilon - u_k^\epsilon) =$$

$$= \frac{\partial u_k^\epsilon(t,x)}{\partial t}, \ \ c_{kj}(x) > 0 \ \text{ for } k \neq j.$$

All possible transition probabilities of the second component as well as the probability that ν_s^ϵ, during a fixed time interval, has no jumps are separated from zero uniformly for $\epsilon \geq 0$ independently of the first component. Therefore, one can expect that the action functional for the first component $X_t^{\epsilon;x,k}$ in $C_{0T}(\mathbf{R}^r)$ as $\epsilon \downarrow 0$ on an absolutely continuous function $\varphi \in C_{0T}(\mathbf{R}^f)$, $\varphi_0 = x$, is given as follows

$$\tilde{S}_{0T}(\varphi) = \inf_{\nu.} \frac{1}{2} \int_0^T |\dot{\varphi}_s - b_{\nu_s}(\varphi_s)|^2 \, ds, \quad \lambda(\epsilon) = \epsilon^{-2},$$

and $\tilde{S}_{0T}(\varphi) = +\infty$ for the rest of $C_{0T}(\mathbf{R}^r)$. This guess is based on (3.3). It is, actually, true, but the functional $\tilde{S}(\varphi)$, contrary to (3.2) and (3.3), is not semi-continuous from below, and therefore it is inconvenient to work with $\tilde{S}_{0T}(\varphi)$. To get the semi-continuous action functional one should consider the semi-continuous version $S_{0T}(\varphi)$ of $\tilde{S}_{0T}(\varphi)$: $S_{0T}(\varphi) = \liminf_{\varphi' \to \varphi} \tilde{S}(\varphi') \wedge \tilde{S}(\varphi)$. It turns out that one can write down a more explicit expression for $S_{0T}(\varphi)$ (see [EF2]):

$$S_{0T} = \begin{cases} \frac{1}{2} \int_0^T R(\varphi_s, \dot{\varphi}_s) \, ds, & \varphi \in C_{0T}, \ \varphi_0 = x, \ \varphi \text{ is absolutely cont.}; \\ +\infty \text{ for the rest of } C_{0T}(\mathbf{R}^r), \end{cases} \tag{3.4}$$

where

$$R(x, v) = \min_{\substack{\alpha_1, \ldots, \alpha_n \geq 0 \\ \sum_1^n \alpha_k = 1}} \left| v - \sum_{k=1}^n \alpha_k b_k(x) \right|^2.$$

One more example: Let $(X_t^\epsilon, \nu_t^\epsilon; P_{x,k})$ be the process corresponding to the system

$$\frac{\partial u_k^\epsilon(t, x)}{\partial t} = \frac{\epsilon^2}{2} \sum_{i,j=1}^r a_k^{ij}(x) \frac{\partial^2 u_k^\epsilon}{\partial x^i \partial x^j} + \sum_{j=1}^n c_{kj}(x)(u_j^\epsilon - u_k^\epsilon).$$

We will see in the following sections that the large deviations for the component X_t^ϵ in $C_{0T}(\mathbf{R}^r)$ as $\epsilon \downarrow 0$ define the wavefront propagation for a class of reaction-diffusion equations. To describe the action functional for X_t^ϵ in $C_{0T}(\mathbf{R}^r)$ as $\epsilon \downarrow 0$, denote by $L(x, v)$ the convex envelope of the paraboloids $\{Z_k = \sum_{i,j=1}^r a_{ij;k}(x)v^i v^j\}$, $k = 1, \ldots, n$:

$$L(x, v) = \min_{\substack{\alpha_1, \ldots, \alpha_n \geq 0 \\ \sum_{k=1}^n \alpha_k = 1}} \sum_{k=1}^n \alpha_k \left(\sum_{i,j=1}^r a_{ij,k}(x)v^i v^j \right).$$

Here $(a_{ij;k}(x)) = (a_k^{ij}(x))^{-1}$, $x \in \mathbf{R}^r$, is a parameter.

The normalized action functional for the component X_t^ϵ, $X_0^\epsilon = x$, $0 \le t \le T$, in $C_{0T}(\mathbf{R}^r)$ as $\epsilon \downarrow 0$ has the form:

$$S_{0T}(\varphi) = \begin{cases} \frac{1}{2} \int_0^T L(\varphi_s, \dot\varphi_s) ds, & \varphi \in C_{0T}(\mathbf{R}^r), \varphi_0 = x, \ \varphi_t \text{ is absolutely cont.}; \\ +\infty \text{ for the rest of } C_{0T}(\mathbf{R}^r). \end{cases}$$

(3.5)

This functional is semi-continuous from below. The normalizing coefficient is ϵ^{-2} (see [F8]).

The calculation of the action functional for a given family of measures is sometimes a difficult problem. A number of approaches to this problem have been accumulated up to the present time (see [FW1], [V1], [Az], [DS]).

For instance, let μ^ϵ be a family of probabilistic measures in $\mathbf{R}^r$ and put

$$H^\epsilon(\alpha) = \ln \int_{\mathbf{R}^r} \exp\{(\alpha, x)\} \mu^\epsilon(dx).$$

The function $H^\epsilon(\alpha)$ is convex, lower semi-continuous, and has values in $(-\infty, +\infty]$ and $H(0) = 0$.

Assume that a function $\lambda(\epsilon)$ exists, $\lim_{\epsilon \downarrow 0} \lambda(\epsilon) = \infty$, such that the limit

$$\lim_{\epsilon \downarrow 0} \lambda^{-1}(\epsilon) H^\epsilon(\lambda(\epsilon)\alpha) = H(\alpha)$$

exists for all $\alpha \in \mathbf{R}^r$. This function is, of course, also convex and $H(0) = 0$. Let us assume that $H(\alpha)$ is semi-continuous, not equal to $-\infty$, and finite in a neighborhood of $\alpha = 0$. Define $L(\beta)$ as Legendre transform of $H(\alpha)$:

$$L(\beta) = \sup_\alpha [(\alpha, \beta) - H(\alpha)].$$

It is clear that $L(\beta)$ is convex.

Suppose additionally that the function $L(\beta)$ is strictly convex at the points of a dense subset of $\{\beta : L(\beta) < \infty\}$. Then the function $L(x)$, $x \in \mathbf{R}^r$, is the normalized action function for the family μ^ϵ as $\epsilon \downarrow 0$ with the normalizing coefficient $\lambda(\epsilon)$. The strict convexity condition for $L(x)$ is fulfilled if the function $H(\alpha)$ is differentiable. A version of this result was first proved by Ju. Gärtner (see the proof in §5.1 of [FW1]).

Using this finite dimensional result, one can calculate the action functional for families of stochastic processes if those processes can be well enough approximated by finite dimensional objects.

Another simple and convenient result concerning calculation of the action functional:

Let S^1 be the normalized action function for the family of probability measures μ^ϵ in a metric space (X_1, ρ_1) with the normalizing coefficient $\lambda(\epsilon)$. Let T

be a continuous mapping of X_1 into a metric space (X_2, ρ_2), and let measures ν^ϵ on X_2 be given by the formula $\nu^\epsilon(A) = \mu^\epsilon(T^{-1}(A))$. Then the normalized action functional $S^2(y)$ for the family ν^ϵ in (X_2, ρ_2) as $\epsilon \downarrow 0$ is given as follows:

$$S^2(y) = \min\{S^1(x) : x \in \varphi^{-1}(y)\},$$

and the normalizing coefficient for ν^ϵ is $\lambda(\epsilon)$ (see §3.3 in [FW1]).

We say that a set $A \subseteq X$ is regular (with respect to the function S) if the infimum of S on the closure $[A]$ of A coincides with the infimum of S on the set (A) of interior points of A:

$$\inf\{S(x), \; x \in [A]\} = \inf\{S(x), \; x \in (A)\}.$$

It is not difficult to derive from the definition that if $S(x)$ is the normalized action function for μ^ϵ in X as $\epsilon \downarrow 0$ and $\lambda(\epsilon)$ is the normalizing coefficient, then for any regular Borel set $A \subseteq X$:

$$\lim_{\epsilon \downarrow 0} \lambda^{-1}(\epsilon) \ln \mu^\epsilon(A) = -\inf\{S(x) : x \in A\}. \tag{3.5}$$

We say in this case that $\mu^\epsilon(A)$ is logarithmically equivalent to $\exp\{-\lambda(\epsilon)\inf\{S(x) : x \in A\}\}$:

$$\mu^\epsilon(X) \asymp \exp\{-\lambda(\epsilon) \cdot \inf\{S(x) : x \in A\}\}, \; \epsilon \downarrow 0.$$

The logarithmic asymptotics of some integrals (expectations) can be expressed through the action function as well: If $F(x)$ is a continuous bounded functional on X, then

$$\int_X \exp\{\lambda(\epsilon)F(x)\}\mu^\epsilon(dx) \asymp \exp\{\lambda(\epsilon) \cdot \sup\{F(x) - S(x) : x \in X\}\} \tag{3.7}$$

as $\epsilon \downarrow 0$.

If $A \subseteq X$ is a Borel set such that

$$\sup\{F(x) - S(x) : x \in (A)\} = \sup\{F(x) - S(x) : x \in [A]\},$$

then one can in (3.7) replace X by A:

$$\int_A \exp\{\lambda(\epsilon)F(x)\}\mu^\epsilon(dx) \asymp \exp\{\lambda(\epsilon) \cdot \sup\{F(x) - S(x) : x \in A\}\}. \tag{3.8}$$

The large deviation theory allows us to obtain a number of interesting asymptotic results concerning second order partial differential equations. For example, consider the fundamental solution (transition density) $p^\epsilon(t, x, y)$ of the parabolic equation

$$\frac{\partial u}{\partial t} = \frac{\epsilon}{2} \sum_{i,j=1}^{r} a^{ij}(x) \frac{\partial^2 u}{\partial x^i \partial x^j}.$$

It was proved by S. R. S. Varadhan, that

$$\lim_{\epsilon \downarrow 0} \epsilon \ln p^\epsilon(t, x, y) = \frac{\rho^2(x, y)}{2t}, \tag{3.9}$$

where $\rho(x, y)$ is the Riemannian metric corresponding to the form $dS^2 = \sum_{i,j=1}^{r} a_{ij}(x) dx^i dx^j$, $(a_{ij}(x)) = (a^{ij}(x))^{-1}$.

Consider now the transition density $p^\epsilon(t, (x, k), (y, \ell))$ of the process $(X_t^\epsilon, \nu_j^\epsilon, P_{x,k})$, corresponding to the system

$$\frac{\partial u_k}{\partial t} = \frac{\epsilon}{2} \sum_{i,j=1}^{r} a_k^{ij}(x) \frac{\partial^2 u_k}{\partial x^i \partial x^j} + \sum_{j=1}^{n} c_{kj}(x)(u_j - u_k), \quad k = 1, \ldots, n.$$

Then one can derive from (3.5) and simple bounds for the density that

$$\lim_{\epsilon \downarrow 0} \epsilon \ln p^\epsilon(t, (x, k), (y, \ell)) = \frac{\tilde{\rho}^2(x, y)}{2t},$$

where $\tilde{\rho}(x, y)$ is a Finsler metric connected with our system.

In general, a Finsler metric $d(x, y)$ in $\mathbb{R}^r$ is defined as follows [R]:

$$d(x, y) = \inf_{\substack{\varphi:[0,1]\to\mathbb{R}^r \\ \varphi_0 = x, \varphi_1 = y}} \int_0^1 h(\varphi_s, \dot{\varphi}_s) \, ds \; ; \; x, y \in \mathbb{R}^r.$$

Here $h(u, v)$, $u, v \in \mathbb{R}^r$, is a non-negative, continuous, function convex in v, such that $h(u, \alpha v) = |\alpha| h(u, v)$ for any real α, and $h(u, v) = 0$ only if $v = 0$. (Actually, the Finsler metric is defined on a smooth manifold and the variable v changes in the tangent space. But in our case the tangent space coincides with $\mathbb{R}^r$.) The homogeneity in v provides independence of the integral, involved in the definition of $d(x, y)$, of the parametrization of the curve φ in $\mathbb{R}^r$. One can easily check that $d(x, y)$ satisfies the usual properties of a metric. If $h(u, v) = \left(\sum_{i,j=1}^{r} a_{ij}(u) v^i v^j \right)^{\frac{1}{2}}$, the metric becomes a Riemannian one.

The function $h(u, v)$ (and the metric $d(x, y)$) can be defined by the family of unit spheres Π_u in the tangent space at each point $u \in \mathbb{R}^r$: $h(u, v) = 1$ if $v \in \Pi_u$; the function $h(u, v)$ is defined for the other values of v by the homogeneity condition $h(u, \alpha v) = |\alpha| h(u, v)$. In the Riemannian case all the unit spheres are ellipsoids. The Finsler metric $\tilde{\rho}(x, y)$ corresponds to the unit spheres Π_x, $x \in \mathbb{R}^r$,

such that Π_x is the convex envelope of the Riemannian spheres $R_x^k = \{v \in \mathbb{R}^r :$
$\sum_{i,j=1}^r a_{ij,k}(x)v^i v^j = 1\}$, $(a_{ij,k}(x)) = \left(a_k^{ij}(x)\right)^{-1}$, where $\left(a_k^{ij}(x)\right)$ is the diffusion
matrix in the k-th equation, $k = 1, \ldots, n$.

An interesting open problem: calculate the pre-exponential factor in the
asymptotic of $p^\epsilon(t,(x,k),(y,\ell))$ as $\epsilon \downarrow 0$. As it is well known, this factor defines
some spectral properties of the system.

The large deviation results for systems allow us to consider the asymptotic
properties of the solutions of the Cauchy problem and mixed problems as it was
done in the single equation case (see §2.3 in [FW1]).

Let us consider in more detail the Dirichlet problem (2.10) with condition
(3.1). First, I will recall some results concerning the case of a single equation
$(n = 1)$:

$$L^\epsilon u^\epsilon(x) = \frac{\epsilon^2}{2} \sum_{i,j=1}^r a^{ij}(x)\frac{\partial^2 u^\epsilon(x)}{\partial x^i \, \partial x^j} + \sum_{i=1}^r b^i(x)\frac{\partial u^\epsilon(x)}{\partial x^i} = 0, \ x \in G \subset \mathbb{R}^r$$

$$\left. u^\epsilon(x)\right|_{\partial G} = \psi(x). \tag{3.10}$$

Assume that all trajectories of the dynamical system (2.2) starting at $x \in G \cup \partial G$
are attracted to a point $0 \in G$ (Fig. 2). Then the typical behavior of a trajectory
of the process (2.1) starting at $x \in G \cup \partial G$ for $\epsilon \ll 1$ will be the following:
trajectory $X_t^{x,\epsilon}$ first goes to a small neighborhood of the equilibrium point 0 and
then makes excursions to the areas distant from 0 again and again coming back
to the neighborhood of 0. The trajectory spends most of the time in a small
neighborhood of the point 0. But, sooner or later, the trajectory $X_t^{x,\epsilon}$ will have
a sufficiently large deviation from 0 such that it hits the boundary ∂G for the
first time. The distribution of the exit point defines the solution $u^\epsilon(x)$. The large
deviation bounds for process $X_t^{x,\epsilon}$ imply that in the generic case there exists,
roughly speaking, one way to leave G that is more probable than all other ways
taken together, if ϵ is small enough. More precisely, introduce the function

$$V(0,x) = V(x) = \inf\{S_{0T}(\varphi) : \varphi \in C_{0T}(\mathbb{R}^r), \ \varphi_0 = 0, \ \varphi_T = x, \ T > 0\}, \tag{3.11}$$

where $S_{0T}(\varphi)$ is the action functional for the family $X_t^{x,\epsilon}$, $0 \le t \le T$, given by
(3.3). It is easy to check that $V(x)$ is a continuous non-negative function vanishing
at the point $x = 0$ only. Suppose that there exists only one point $x_0 \in \partial G$ such
that

$$V(x_0) = \min_{x \in \partial G} V(x) = V_0.$$

Then, using the large deviation estimates, one can prove that first exit from G
occurs in a small neighborhood of the point x_0 with probability close to 1 as

$\epsilon \ll 1$: for any $\delta > 0$ and $x \in G$

$$\lim_{\epsilon \downarrow 0} P_x\{|\, X^\epsilon_{\tau^\epsilon} - x_0 \,| > \delta\} = 0,$$

where $\tau^\epsilon = \min\{t : X^\epsilon_t \notin G\}$.

Now, taking into account the probabilistic representation of the solution of problem (3.10), we have:

$$u^\epsilon(x) = E_x \psi(X^\epsilon_{\tau^\epsilon}) \to \psi(x_0), \quad \text{as } \epsilon \to 0.$$

This result can be found in [FW1]. The case of several equilibrium points or other attractors inside G is also considered there.

Before switching to the Dirichlet problem for systems with condition (3.1), I want to make a remark: Stationary problems for PDE's arise as a rule as a result of stabilization of solutions of corresponding evolutionary problem as time tends to infinity. For example, problem (3.10) describes the limit as $t \to \infty$ of the solution of the mixed problem

$$\frac{\partial u^\epsilon(t,x)}{\partial t} = L^\epsilon u^\epsilon(t,x), \quad t > 0, \ x \in G,$$

$$u^\epsilon(t,x)\Big|_{\partial G} = \psi(x), \quad u^\epsilon(0,x) = g(x). \tag{3.12}$$

If a small parameter ϵ is included in the evolutionary equation, then one should study the two parameter asymptotic problem: ϵ^{-1}, $t \to \infty$. The limit, in general, depends on how the point (ϵ^{-1}, t) approaches infinity. Consideration of the behavior of the solution of problem (3.10) as $\epsilon \downarrow 0$ corresponds to the case when t tends to infinity much faster than ϵ^{-1}. If the Levinson conditions are fulfilled, it is easy to see that the limit of the solution of (3.12) as ϵ^{-1}, $t \to \infty$ is independent of the relation between ϵ^{-1} and t. But the situation is different in the large deviation case. For example, if, as it shows in Fig. 2, the point 0 is the only attractor of the dynamical system (2.2) in $G \cup \partial G$ and the initial function $g(x)$ in (3.13) is continuous, then $u^\epsilon(t,x) \to g(0)$ as ϵ^{-1}, $t \to \infty$ and $\overline{\lim}_{\epsilon, t \to \infty} \epsilon \ln t < V_0$, and $u^\epsilon(t,x) \to \psi(X_0)$ as ϵ^{-1}, $t \to \infty$, so that

$$\underline{\lim}_{\epsilon^{-1}, \ t \to \infty} \epsilon \ln t > V_0.$$

It is connected with the fact that for any $h > 0$

$$\lim_{\epsilon \to 0} P_x\left\{ e^{\frac{V_0 - h}{\epsilon}} < \tau^\epsilon < e^{\frac{V_0 + h}{\epsilon}} \right\} = 1,$$

(See Ch. 4 in [FW1]).

If the dynamical system has many attractors, the situation becomes more complicated: one can introduce a hierarchy of cycles, each cycle has its own characteristic time, and the main state, $u^\epsilon(t,x)$ has different limits as ϵ^{-1}, $t \to \infty$

depending on the behavior $\epsilon \ln t(\epsilon)$. These problems were considered in [F3], See also [FW1].

Let us consider now the Dirichlet problem for a system:

$$\frac{\epsilon^2}{2}\Delta u_k^\epsilon(x) + \sum_{i=1}^r b_k^i(x)\frac{\partial u_k^\epsilon(x)}{\partial x^i} + \sum_{j=1}^n c_{kj}(x)(u_j^\epsilon(x) - u_k^\epsilon(x)) = 0,$$

$$x \in G, \ u_k^\epsilon(x)\Big|_{\partial G} = \psi_k(x), \ k = 1, 2, \ldots, n. \tag{3.13}$$

We assume that $c_{kj}(x) > 0$ and that condition (3.1) is fulfilled.

The Markov process $(X_t^\epsilon, \nu_t^\epsilon, P_{x,k})$ in the phase space $R^r \times \{1, \ldots, n\}$, corresponding to (3.13) is defined as follows:

$$\dot{X}_t^\epsilon = b_{\nu_t^\epsilon}(X_t^\epsilon) + \epsilon \dot{W}_t,$$

$$P_{x,k}\{\nu_{t+\Delta}^\epsilon = \ell\} = c_{k\ell}(x)\Delta + o(\Delta), \ \Delta \downarrow 0.$$

The solution of problem (3.13) can be written in the form

$$u_k^\epsilon(x) = E_{x,k}\psi_{\nu_{\tau^\epsilon}^\epsilon}(X_{\tau^\epsilon}^\epsilon), \tag{3.14}$$

where τ^ϵ is the exit time:

$$\tau^\epsilon = \min\{t : X_t^\epsilon \notin G\}.$$

The normalized action functional $S_{0T}(\varphi)$ for the family X_t^ϵ in $C_{0T}(R^r)$ as $\epsilon \downarrow 0$ is given by equality (3.4). Define

$$V(x, y) = \inf\{S_{0T}(\varphi), \ \varphi \in C_{0T}, \ \varphi_0 = x, \ \varphi_T = y, \ T > 0\}. \tag{3.15}$$

It is easy to check that $V(x, y)$ is continuous and non-negative. It follows from (3.1) that $V(x, y) > 0$ if $x \in G$ and $y \in \partial G$.

We say that $x \sim y$ if

$$V(x, y) = V(y, x) = 0.$$

The equivalence of x and y means, roughly speaking, that the transition from one point to any neighborhood of the other is not due to the large deviations. As I mentioned above, condition (3.1) means that an interior point of G cannot be equivalent to any point of the boundary.

The next assumption is the counterpart of the assumption made in the single equation case, that the field $b(x)$ has inside G a unique attracting point. Recall, that a set $A \subset R^r$ is called an $\omega - limit$ set for a trajectory X_t of a dynamical system in R^r, if $A = \{z \in R^r : \liminf_{t\to\infty} | X_t - z |= 0\}$.

Assumption 3.1 *Let a compact set $K \subset G$ exist such that any two points of K are equivalent, and that the ω-limit sets of trajectories of dynamical systems*

$$\dot{X}_t^{(k)} = b_k(X_t^{(k)}), \ \ k = 1, \ldots, n,$$

starting in $G \cup \partial G$ belong to K.

Remark.   Assumption 3.1 can be relaxed: we need, actually, the fact that the trajectories of the degenerate process $(X_t^0, \nu_t^0, P_{x,k})$ enter the neighborhood of the compact K with probability 1, and that K consists of equivalent points.

Assumption 3.2 *There exists $x_0 \in \partial G$ such that*

$$V(x, x_0) < V(x, y)$$

for any $x \in K$ and $y \in \partial G$, $y \neq x_0$.

This assumption is a generalization of the assumption that the function $V(x)$ defined by (3.11) has a unique minimum on the boundary.

And, finally, we will make one more assumption that has no counterpart in the single equation case.

Assumption 3.3 *There exists i_0, $1 \leq i_0 \leq n$, such that at the point $x_0 \in \partial G$, defined in Assumption 3.2, the following inequalities hold:*

$$(b_{i_0}(x_0), n(x_0)) > (b_i(x_0), n(x_0)), \ \ 1 \leq i \leq n, \ i \neq i_0.$$

Note that Assumptions 3.2 and 3.3 are satisfied in the generic case.

Theorem 3.1 *Let (3.1) and Assumptions 3.1 and 3.2 be fulfilled. Let $\tau^\epsilon = \min\{t :$ $X_t^\epsilon \notin G\}$. Then*

$$\lim_{\epsilon \to 0} P_{x,k}\{| X^\epsilon(\tau^\epsilon) - x_0 | > \delta\} = 0$$

for any $\delta > 0$, $1 \leq k \leq n$, uniformly in $x \in F$ for any compact $F \subset G$.

If, in addition, Assumption 3.3 is satisfied, then

$$\lim_{\epsilon \to 0} P_{x,k}\{\nu^\epsilon(\tau^\epsilon) = i_0\} = 1$$

for $1 \leq k \leq n$ and $x \in K \subset G$.

Corollary.   Let (3.1) and Assumptions 3.1-3.3 hold. Then

$$\lim_{\epsilon \downarrow 0} u_k^\epsilon(x) = \psi_{i_0}(x_0), \ \ 1 \leq k \leq n,$$

uniformly in $x \in F \subset G$, where $u_k^\epsilon(x)$ is the solution of the problem (3.13).

The proof of the Corollary easily follows from the Theorem 3.1 and the representation

$$u^\epsilon(x) = E_x \psi_{\nu^\epsilon_{\tau^\epsilon}}(X^\epsilon_{\tau^\epsilon}).$$

Let us now outline the proof of Theorem 3.1. The full proof can be found in [EF2].

Let $\mathcal{E}_\delta$ be the δ-neighborhood of the compact set K, $\gamma_\delta = \partial\mathcal{E}_\delta$. The trajectories X^ϵ_t, starting from any $x \in G$, $k \in \{1,\dots,n\}$, hit γ_δ before ∂G with probability close to 1 as ϵ is small enough. This follows from (3.1) and Assumption 3.1. Thus, taking into account the strong Markov property of the process $(X^\epsilon_t, \nu^\epsilon_t; P_{x,k})$, it is sufficient to prove Theorem 3.1 for $x \in \gamma_\delta$, $k \in \{1,\dots,n\}$.

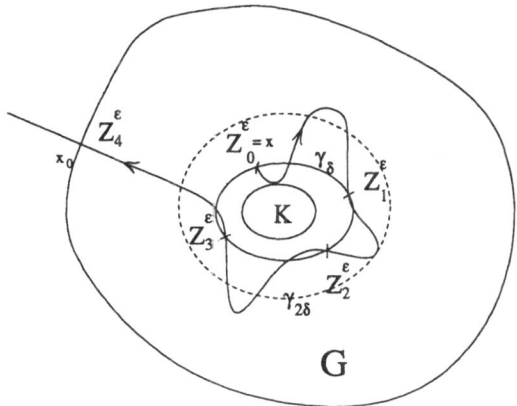

Figure 4

First, define Markov times $\sigma_0 < \tau_1 < \sigma_1 < \cdots < \tau_n < \sigma_n \cdots$ as follows (see Fig. 4):

$$\sigma_0 = \min\{t > 0,\ X^\epsilon_t \in \gamma_{2\delta}\},$$
$$\tau_1 = \min\{t > \sigma_0 : X^\epsilon_t \in \gamma_\delta \cup \partial G\},$$
$$\sigma_1 = \min\{t > \tau_1 : X^\epsilon_t \in \gamma_{2\delta}\}, \cdots$$

$$\tau_{n+1} = \min\{t > \sigma_n : x^\epsilon_t \in \gamma_\delta \cup \partial G\},$$
$$\sigma_{n+1} = \min\{t > \tau_{n+1} : X^\epsilon_t \in \gamma_{2\delta}\}, \cdots$$

Now we define a Markov chain $(Z^\epsilon_n, \hat\nu^\epsilon_n)$ in the phase space $\{\gamma_\delta \cup \partial G\} \times \{1,\dots,n\}$: $Z^\epsilon_n = X^\epsilon_{\tau_n}, \hat\nu^\epsilon_n = \nu^\epsilon_{\tau_n}$. The first exit of X^ϵ_t from G occurs when the component Z^ϵ_n of the chain first time belongs to ∂G. Using the large deviation estimates for the processes $(X^\epsilon_t, \nu^\epsilon_t; P_{x,k})$ as $\epsilon \to 0$, one can prove in a standard way (see Ch. 4 in [FW1]), that Z^ϵ_n, starting from any $x \in \gamma_\delta$ and $k \in \{1,\dots,n\}$ reaches ∂G for the first time in a small neighborhood of the point $x_0 \in \partial G$, introduced in Assumption 3.2, with probability close to 1 as ϵ and δ are small enough. This implies the first statement of Theorem 3.1.

To prove the second statement, one should use the fact that the extremals of the variational problem

$$\inf\{S_{0T}(\varphi), \ \varphi_0 \in K, \ \varphi_T \in \partial G, \ T > 0\}$$

spend in δ-neighborhood $\mathcal{E}_\delta = \{x \in G : \rho(x, \partial G) < \delta\}$ of ∂G a time of order δ as $\delta \downarrow 0$. Therefore with probability close to 1, as δ is small, the second component ν_t^ϵ has no jumps during this time, and X_t^ϵ hits the boundary for the value of the second coordinate ν_t^ϵ such that the transition of X_t^ϵ from $\partial \mathcal{E} \backslash \partial G$ to ∂G is "the easiest." Simple estimates show that this transition is the easiest when the second component is equal to i_0 defined in the Assumption 3.3.

The situation here is similar to a problem considered in [GF1] for one equation. The detailed proof in the case of systems can be found in [EF2].

Remark. The result of Theorem 3.1 is independent of the coefficients $c_{ij}(x)$. Only one assumption concerning $c_{ij}(x)$ is important: The matrix $(c_{ij}(x))$ should be ergodic for any $x \in \mathbb{R}^r$.

In conclusion of this section consider one more way to introduce a small parameter in the linear system.

Consider the following Dirichlet problem (compare (2.30)):

$$L_k^\epsilon u_k^\epsilon(x) + \frac{1}{\epsilon} \sum c_{kj}(x)(u_j^\epsilon(x) - u_k^\epsilon(x)) = 0, \ x \in G,$$

$$u_k^\epsilon(x)\Big|_{\partial G} = g_k(x), \ k \in \{1, \ldots, n\}, \tag{3.16}$$

where

$$L_k^\epsilon = \frac{\epsilon}{2} \sum_{i,j=1}^r a_k^{ij}(x)\frac{\partial^2}{\partial x^i \partial x^j} + \sum_{i=1}^r b_k^i(x)\frac{\partial}{\partial x^i}.$$

As usual, we assume that the coefficients are Lipschitz continuous and the forms $\sum_{i,j=1}^r a_k^{ij}(x)\lambda_i\lambda_j$, $k = 1, \ldots, n$, are positively defined, $c_{kj}(x) > 0$.

Now we have a large parameter ϵ^{-1} in the terms governing the jumps of the second component of the process $(X_t^\epsilon, \nu_t^\epsilon; P_{x,k})$ corresponding to system (3.16). Therefore, before the first component makes a small displacement from a starting point $x \in \mathbb{R}^r$, the second component makes many transitions with intensities $\approx \epsilon^{-1}c_{ij}(x)$ if $\epsilon \ll 1$. The distribution of the second component approaches the stationary distribution $(\gamma_1(x), \gamma_2(x), \cdots, \gamma_n(x))$ of the Markov chain with intensities $c_{ij}(x)$, when x is fixed. Taking into account that the diffusion coefficients are also of order ϵ as $\epsilon \downarrow 0$, one can conclude that the first component X_t^ϵ approaches the trajectory of the averaged dynamical system

$$\dot{\bar{X}}_t = \bar{b}(\bar{X}_t), \ \bar{b}(x) = \sum_{k=1}^n \gamma_k(x)b_k(x), \tag{3.17}$$

as $\epsilon \downarrow 0$ with the same initial condition $\bar{X}_0 = X_0^\epsilon = x$.

The component X_t^ϵ can be considered as a random perturbation of the dynamical system (3.17):

$$\dot{X}_t^{\epsilon,x} = b_{\nu_t^\epsilon}(X_t^{\epsilon,x}) + \sqrt{\epsilon}\sigma_{\nu_t^\epsilon}(X_t^{\epsilon,x})\dot{W}_t, \ \ X_0^{\epsilon,x} = x.$$

The trajectories of system (3.17) play the same part for problem (3.16) as process $(X_t^0, \nu_t^0, P_{x,k})$ for system (3.13) or the dynamical system $\dot{X}_t = b(X_t)$ for problem (3.10).

If the trajectories of (3.17) leave the domain G in a finite time and cross ∂G in a regular way, we have the Levinson case. If $\bar{b}(x)$ is directed inside G on the whole boundary ∂G we have the large deviation case.

Deviations of X_t^ϵ from $\bar{X}_t$ occur due to two factors: first, due to the small diffusion and, second, due to the deviations of the occupation times of the second component from the invariant distribution. The probabilities of both of these deviations have the same logarithmic order as $\epsilon \downarrow 0$. Thus we should take into account both of them.

Define the occupation times for the second component:

$$Z_t^\epsilon = \left(\int_0^t \chi_1(\nu_s^\epsilon)\,ds, \ldots, \int_0^t \chi_n(\nu_s^\epsilon)\,ds \right),$$

where $\chi_k(i), = \ i \in \{1,\ldots,n\}$, is the indicator function of the state k. Consider the matrix $(A_{km}(x,p,\alpha))$:

$$A_{km}(x,p,\alpha) = (\hat{c}_{km}(x) + \Lambda), \ \ x \in \mathrm{R}^r, \ p, \alpha \in \mathrm{R}^r,$$

where $\hat{c}_{km}(x) = c_{km}(x)$ for $k \neq m$ and $\hat{c}_{kk}(x) = - \sum_{j:j\neq k} c_{kj}(x); \ \Lambda(x,p,\alpha)$ is the diagonal matrix,

$$\Lambda_k(x,p,\alpha) = \frac{1}{2}(p \cdot a_k(x)p) + (b_k(x) \cdot p) + \alpha_k.$$

Denote by $\lambda(x,p,\alpha)$ the eigenvalue of the matrix $(A_{km}(x,p,\alpha))$ corresponding to the positive eigenvector. Such a $\lambda(x,p,\alpha)$ exists and is convex in $p, \alpha \in \mathrm{R}^n$.

Denote by $\eta(x,q,\beta)$ the Legendre transformation of $\lambda(x,p,\alpha)$ in (p,α):

$$\eta(x,q,\beta) = \sup_{p\in\mathrm{R}^r, \alpha\in\mathrm{R}^n} ((q \cdot p) + (\alpha \cdot \beta) - \lambda(x,p,\alpha)),$$

$$x \in \mathrm{R}^r, \ q \in \mathrm{R}^r, \ \beta \in \mathrm{R}^n.$$

Let $C_{0T}(\mathrm{R}^r)$ be the space of continuous functions on [0,T] with values in R^r,

$$C_{0T}^+(\mathrm{R}^n) = \{\mu. = (\mu_1,\ldots,\mu_n) : [0,T] \to \mathrm{R}^n,$$

$$\mu_i(0) = 0, \ \mu_i(t) \text{ is non-decreasing and } \sum \mu_i(t) \equiv t\}.$$

Define

$$
S_{0T}(\varphi) = \begin{cases} \int_0^t \eta(\varphi_s, \dot{\varphi}_s, \dot{\mu}_s)\, ds, & \text{if } \varphi \in C_{0T}(\mathbf{R}^r),\, \varphi_0 = x, \text{ and } \mu \in C_{0T}^+(\mathbf{R}^n) \\ \qquad\quad \text{are absolutely continuous,} \\[2mm] +\infty, & \text{otherwise.} \end{cases}
$$

$$(3.18)$$

One can prove that $\epsilon^{-1} S_{0T}(\varphi, \mu)$ is the action functional for the family $(X_t^\epsilon, Z_t^\epsilon)$ in the uniform topology as $\epsilon \downarrow 0$ (see [FL1]). We will use the action functional for $(X_t^\epsilon, Z_t^\epsilon)$ later when we consider wavefronts for a class of semilinear systems. Here we need only a part of this result: the action functional for the processes X_t^ϵ as $\epsilon \downarrow 0$.

Define

$$
\rho(x, q) = \inf_{\beta \in \mathbf{R}^n} \eta(x, q, \beta).
$$

It is easy to check that $\rho(x, q)$ is the Legendre transformation of $\lambda(x, p, 0)$ in p. Using the contraction principle one can derive that the functional

$$
I_{0T}(\varphi) = \begin{cases} \int_0^T \rho(\varphi_s, \dot{\varphi}_s)\, ds, & \text{if } \varphi \in C_{0T}(\mathbf{R}^r) \text{ is absolutely continuous, } \varphi_0 = x, \\[2mm] +\infty, & \text{otherwise} \end{cases}
$$

is the normalized action functional for the family X_t^ϵ, $X_0^\epsilon = x$, in $C_{0T}(\mathbf{R}^r)$ as $\epsilon \downarrow 0$ with the normalizing coefficient ϵ^{-1}. Moreover, it can be easily checked that $I_{0T}(\varphi)$ is equal to zero if and only if φ_s is the trajectory of the averaged system (3.17) starting at $x \in \mathbf{R}^r$. In particular, the uniform convergence of X_t^ϵ to $\bar{X}_t$, $t \in [0, T]$, in probability follows from the last statement. Using this convergence one can prove that in the Levinson case, that is, when $\bar{X}_t$ leaves G in a finite time and crosses ∂G in the regular way, X_t^ϵ exits the domain G near the point where $\bar{X}_t$, $\bar{X}_0 = X_0^\epsilon = x$, leaves the domain. But to calculate the limit of $u_k^\epsilon(x)$ we need to know not only the limit of $X_{\tau^\epsilon}^\epsilon$ but of $\nu_{\tau^\epsilon}^\epsilon$ as well; τ^ϵ is, as usual, the exit time: $\tau^\epsilon = \min\{t : X_t^\epsilon \notin G\}$.

A point $y \in \partial G$ is called regular if $\bar{b}(y)$ points toward the outside of G. One could expect, that, if the Levinson condition is fulfilled, then $\lim_{\epsilon \downarrow 0} u_k^\epsilon(x)$ is determined only by the values of the boundary functions $g_k(y)$ at the regular points $y \in \partial G$ and k such that $b_k(y)$ is directed outward. But it turns out that it is not the case: The interplay between the small diffusion and fast jumps of the ν-component leads to the situation when all $g_k(y)$, $k = 1, \ldots, n$, will influence the limit. This limit in the Levinson case is found in [FL1].

If the vector field $\bar{b}(y)$ on ∂G is directed inside G, that is $(n(y) \cdot b(y)) > 0$, $y \in \partial G$, where $n(y)$ is the inward normal to ∂G, we have the large deviation case. As usual, in the generic large deviation case with one attracting set, a point $y_0 \in \partial G$ exist such that $X_{\tau^\epsilon}^\epsilon \to y_0$ in probability as $\epsilon \downarrow 0$. The point y_0 is defined

by a quasipotential corresponding to the problem. The quasipotential is defined
by the action functional in the standard way (compare [FW1] Ch. 4). To calculate
the limiting distribution of $\nu^\epsilon_{\tau\epsilon}$ as $\epsilon \downarrow 0$ one needs some special considerations. This
was done in [FL1]. This limiting distribution together with the point y_0 define the
limit of $u^\epsilon_k(x)$ as $\epsilon \downarrow 0$.

The problem (3.16) is still interesting if all diffusion coefficients vanish. In
that case the boundary conditions should be posed not on the whole geometric
boundary of the domain, but only on the regular part of it. Consider the equations

$$b_k(x) \cdot \nabla u^\epsilon_k(x) + \frac{1}{\epsilon} \sum c_{kj}(x) \left(u^\epsilon_j(x) - u^\epsilon_k(x) \right) = 0 \quad , x \in G, \quad k = 1, \ldots n. \quad (3.19)$$

Here $c_{kj}(x) > 0$, $b_k(x) = \left(b^1_k(x), \ldots, b^r_k(x) \right)$, $0 < \epsilon \ll 1$. Let $(X^\epsilon_t, \nu^\epsilon_t; P_{x,k})$ be the
process corresponding to this system. Assume that the weak Levinson condition
(see Section 2) is fulfilled. Let us add the Dirichlet boundary conditions:

$$u^\epsilon_k(x)|_{\Gamma^+_k} = g_k(x), \quad k = 1, \ldots, n. \quad (3.20)$$

The regular parts Γ^+_k of δG are defined in Section 2. Problem (3.19)–(3.20) has
a unique solution for any $\epsilon > 0$. Let $\bar{b}(x)$ be the averaged vector field defined in
(3.17). Consider equation

$$\bar{b}(x) \cdot \nabla \bar{u}(x) = 0, \quad x \in G, \quad (3.21)$$

and the dynamical system $\dot{\bar{X}}^x_t = \bar{b} \left(\bar{X}^x_t \right)$, $\bar{X}^x_0 = x$.

Again, there are three extrem cases in the asymptotic behavior of the solutions of problem (3.19)–(3.20). First, when the Levison condition is fulfilled for
(3.21). This case is similar to the Levinson case for one equation mentioned in section 2 (See also [F6], Ch.4). The limiting function is independent of k and satisfies
(3.21). But there is an additional difficulty concerning the boundary conditions.
If for each point $x \in \delta G$ there exits not more than one vector field $b_k(x)$ directed
outside G, say $b_{k^*(x)}$, and all other fields are directed strictly inside, then, under
mild conditions, the boundary function for the limiting function is $g_{k^*(x)}(x)$. The
boundary conditions should be posed only at the part of the geometric boundary
which is regular for $\bar{b}(x)$.

We face the second case, if $\bar{b}(x)$ on ∂G is directed strictly inside G This is
the large deviation case (see [FL1]).

And, finally, the third case, when the field $\bar{b}(x)$ does not hinder the exit but
does not help to exit G as well. We have a special example of the last case if $\bar{b}(x) \equiv$
0. Then, as it follows from [Kh1], $X^\epsilon_{t/\epsilon}$ converges weakly to a diffusion process. The
corresponding generator $\bar{L}$ can be expressed through the coefficients of equation
(3.19). The limits $\lim_{\epsilon \downarrow 0} u^\epsilon_k(x) = u(x)$ are independent of k and $\bar{L}u(x) = 0$, $x \in G$.
The calculation of the boundary conditions is a more delicate question. The third
case is closely connected with problems studied in the next two sections.

4 Averaging Principle for Stochastic Processes and for Partial Differential Equations

Consider a dynamical system in R^2:

$$\dot{X}_t = b(X_t), \quad X_0 = x \in R^2 \tag{4.1}$$

Assume that system (4.1) has a first integral $H(x)$ and that this function is of C^2-class: $H(X_t) \stackrel{t}{=} H(x)$. Let the function $H(x)$ have just one critical point – a minimum at the origin 0, $H(0) = 0$, $H(x) > 0$ for $x \neq 0$, and suppose that all the level sets $C(y) = \{x \in R^2 : H(x) = y\}$, $y \geq 0$, are compact. Moreover, let $b(y) \neq 0$ for $y \neq 0$. The corresponding phase picture is given in Fig. 5.

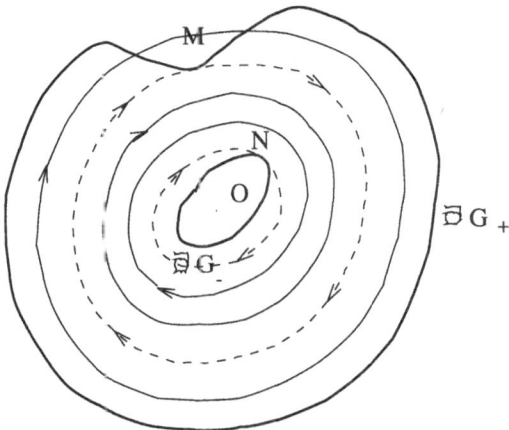

Figure 5

On each separate level set $C(y)$, $y > 0$, the trajectory is periodic with period

$$T(y) = \int\limits_{C(y)} \frac{dl}{|b(y)|},$$

where dl is the length element on $C(y)$. The dynamical system has on each level set $C(y)$ a unique invariant measure. The normalized density of this measure has the form:

$$M_y(x) = \frac{1}{T(y)} \frac{1}{|b(x)|}$$

Consider now small white noise type perturbations of this dynamical system:

$$\dot{\tilde{X}}_t^\epsilon = b(\tilde{X}_t^\epsilon) + \epsilon\sigma(\tilde{X}_t^\epsilon)\dot{W}_t, \quad \tilde{X}_0^\epsilon = x \in R^2 \tag{4.2}$$

If we consider the process $\tilde{X}_t^\epsilon$ on a finite time interval $[0, T]$, then, of course, $\tilde{X}_t^\epsilon \to X_t$ uniformly on $[0, T]$ in probability as $\epsilon \downarrow 0$. One can prove that the normalized deviations $\zeta_t^\epsilon = \frac{1}{\epsilon}(\tilde{X}_t^\epsilon - X_t)$, $0 \le t \le T < \infty$, converge weakly as $\epsilon \downarrow 0$ in the space of continuous functions C_{0T} to a Markov Gaussian process. But, as a rule, the behavior of the perturbed system on large time intervals is of interest. The noise on such intervals may cause essential deviations from the dynamical system. For example, the trajectory X_t^ϵ, starting at a point x can deviate from the level set $C_{H(x)}$ on a distance of order 1 as $\epsilon \downarrow 0$. Note that all points of the phase space for such a system are equivalent in the sense of Section 3: $V(x, y) = \inf\{S_{0T}(\varphi), \varphi_0 = x, \varphi_T = y, T \ge 0\} = 0$ for any $x, y \in \mathrm{R}^2$, where the action functional $S_{0T}(\varphi)$ is defined by (3.3). This, roughly speaking, means that the transition from one level to another is not a large deviation. One can see that, for the system described above, time intervals of order ϵ^{-2} should be considered in order to observe such an event with a probability of order 1 as $\epsilon \downarrow 0$. Therefore, to deal with finite time intervals, let us rescale time: consider $X_t^\epsilon = \tilde{X}_{t/e^2}^\epsilon$. Then X_t^ϵ satisfies the equation:

$$\dot{X}_t^\epsilon = \frac{1}{\epsilon^2} b(X_t^\epsilon) + \sigma(X_t^\epsilon) \dot{W}_t, \quad X_0^\epsilon = x. \tag{4.3}$$

The motion described by (4.3) has two components: one along the periodic trajectories of the dynamical system, and the other is a displacement in the direction transversal to the trajectories of (4.1). The first motion is fast: its speed is, in a sense, of order ϵ^{-2} as $\epsilon \downarrow 0$. The second component is relatively slow: the mean value of the displacement in the transversal direction during a finite time interval is of order 1 as $\epsilon \downarrow 0$.

One can expect that the distribution of the fast component at time t will be close, for $\epsilon \ll 1$, to the invariant distribution of the system (4.1) on the level set C_y, where y is defined by the slow component. So that, the main problem is to describe the slow evolution.

From the PDE point of view our problem consists of studying various problems for the family of operators

$$L^\epsilon = \frac{\epsilon^2}{2} \sum_{i,j=1}^{2} a^{ij}(x) \frac{\partial^2}{\partial x^i \partial x^j} + \sum_{i=1}^{2} b^i(x) \frac{\partial}{\partial x^i}, \quad 0 < \epsilon \ll 1,$$

when the field $b(x) = (b^1(x), b^2(x))$ behaves as in Fig. 5. For example, the Dirichlet problem is of interest:

$$L^\epsilon u^\epsilon(x) = 0, \ x \in G \subset \mathrm{R}^2, \ u^\epsilon(x)\big|_{\partial G} = \psi(x) \tag{4.4}$$

Let the domain G be as in Fig. 5: the boundary ∂G consists of two smooth curves ∂G_- and ∂G_+; M is the (unique) point where the level set $C(y_{\max})$, $y_{\max} = \max\{y : C(y) \in G \cup \partial G\}$ touches ∂G_+; N is the point where $C(y_{\min})$, $y_{\min} =$

$\min\{y : C(y) \in G \cup \partial G\}$, touches ∂G_-. The level sets $C(y_{\max})$ and $C'(y_{\min})$ are shown in Fig. 5 by dotted lines. Note that in the case under consideration the drift $b(x)$ does not help a particle (at least if X_t^ϵ starts at a point x with $y_{\min} < H(x) < y_{\max}$) to leave the domain like in the Levinson case, but at the same time it does not hinder the exit like in the large deviation case.

It is convenient to describe the slow component by the value of the first integral $H(x)$. Applying the Itô formula to $H(X_t^\epsilon)$ we obtain :

$$
\begin{aligned}
H(X_t^\epsilon) - H(x) = & \int_0^t (\nabla H(X_s^\epsilon) \cdot \sigma(X_s^\epsilon)\, dW_s) \\
& + \frac{1}{\epsilon^2} \int_0^t (\nabla H(X_s^\epsilon) \cdot b(X_s^\epsilon))\, ds + \frac{1}{2} \int_o^t \sum_{i,j=1}^r a^{ij}(X_s^\epsilon) \frac{\partial^2 H(X_s^\epsilon)}{\partial x^i\, \partial x^j}\, ds
\end{aligned}
\tag{4.5}
$$

Since $H(x)$ is a smooth first integral

$$
\nabla H(x) \cdot b(x) = 0,
$$

and the term with the factor $1/\epsilon^2$ in (4.5) disappears.

Before the diffusing particle makes a small displacement transversally to the level sets directions it makes many rotations along the periodic trajectories. So the drift and diffusion coefficients for the slow motion will be averaged along corresponding level sets and will asymptotically depend not on the position of the particle but just on the level sets i.e. on the value of the first integral $H(x)$. The averaging should be made in accordance with the time spent by the trajectory X_t^ϵ in the various parts of the level set. This time is asymptotically, as $\epsilon \downarrow 0$, proportional to the invariant measure of the dynamical system on that level set.

This gives us the following expression for the drift in the limiting slow motion at a level $H = y$:

$$
\bar{B}(y) = \frac{1}{2T(y)} \int_{C(y)} \frac{1}{|b(x)|} \sum_{i,j=1}^2 a^{ij}(x) \frac{\partial^2 H(x)}{\partial x^i\, \partial x^j}\, dl,
$$

How one should average the diffusion coefficient? Using the self-similarity property of the Wiener process, one can write:

$$
\begin{aligned}
\int_0^h \sigma(X_s^\epsilon)\, dW_s \cdot \nabla H(X_s^\epsilon) = & \int_0^h \sum_{i,j=1}^2 \sigma_i^j(X_s^\epsilon) \frac{\partial H}{\partial x^i}(X_s^\epsilon)\, dW_s^j = \\
& = \widetilde{W}\left(\int_0^h \sum_{i,j=1}^2 a^{ij}(X_s^\epsilon) \frac{\partial H}{\partial x^i}(X_s^\epsilon) \frac{\partial H}{\partial x^j}(X_s^\epsilon)\, ds \right),
\end{aligned}
$$

where $\widetilde{W}_t$ is an one-dimensional Wiener process, $a(x) = (a^{ij}(x)) = \sigma(x)\sigma^*(x)$.

Now we already have an integral in ds not in dW_s, and we can use the averaging procedure:

$$\int_0^h \sum_{i,j=1}^2 a^{ij}(X_s^\epsilon) \frac{\partial H}{\partial x^i}(X_s^\epsilon) \frac{\partial H}{\partial x^j}(X_s^\epsilon) \, ds$$
$$\approx \frac{h}{T(y)} \int_{C(y)} \frac{(a(x)\nabla H(x)\cdot\nabla H(x)) \, dl}{|b(x)|} + o(h),$$

when $\epsilon \ll 1$, h is small and $X_o^\epsilon = x$ is such that $H(x) = y$. It means that the diffusion coefficient for the limiting (as $\epsilon \downarrow 0$) slow motion is equal to

$$\bar{\sigma}^2(y) = \frac{1}{T(y)} \int_{C(y)} \frac{a(x)\nabla H(x) \cdot \nabla H(x)}{|b(x)|} \, dl$$

One can prove that the processes $Y_t^\epsilon = H(X_t^\epsilon)$, $0 \leq t \leq T < \infty$, converge weakly in C_{0T} as $\epsilon \downarrow 0$ to the diffusion process governed by the operator

$$\bar{L} = \frac{1}{2}\bar{\sigma}^2(y)\frac{d^2}{dy^2} + \bar{b}(y)\frac{d}{dy}$$

This process, of course, can be described by the stochastic differential equation

$$dY_t = \bar{\sigma}(Y_t), d\tilde{W}_t + \bar{b}(Y_t) \, dt.$$

To prove this weak convergence one should, first, check that the family $\{Y_t^\epsilon, \ 0 \leq t \leq T\}$ is tight in C_{0T}. This can be easily done using (4.5), since the coefficients in (4.5) are bounded. Then, using the above mentioned arguments, one can prove convergence of the finite-dimensional distributions of $Y_t^\epsilon = H(X_t^\epsilon)$ to correspondimg distributions of the process Y_t. It was first done by Khasmin-skii[Kh2] (see also [F6]).

Consider an example. Let (4.1) be a Hamiltonian system in $\mathbb{R}^2$:

$$\dot{X}_t = \bar{\nabla}H(X_t), \quad \bar{\nabla}H(x) = \left(\frac{\partial H(x)}{\partial x^2}, -\frac{\partial H(x)}{\partial x^1}\right) \tag{4.6}$$

We assume that the Hamiltonian $H(x)$ is smooth and has the properties mentioned before in this section. It is clear that $H(x)$ is a first integral for system (4.5):

$$\frac{dH(X_t)}{dt} = \nabla H(X_t) \cdot \bar{\nabla}H(X_t) \equiv 0$$

Let $\sigma(x)$ be the unit matrix, so that

$$\dot{X}_t^\epsilon = \frac{1}{\epsilon^2}\bar{\nabla}H(X_t^\epsilon) + \dot{W}_t$$

Then the diffusion and drift coefficients for the limiting slow motion are the following:

$$\bar{b}(y) = \frac{1}{2T(y)} \int_{C(y)} \frac{\Delta H(x)\, dl}{|\nabla H(x)|}, \quad \bar{\sigma}^2(y) = \frac{1}{T(y)} \int_{C(y)} |\nabla H(x)|\, dl,$$

$$T(y) = \int_{C(y)} \frac{dl}{|\nabla H(x)|}$$

What does this result give for PDE's?

Theorem 4.1 *Let $u^\epsilon(x)$ be the solution of problem (4.4). Assume that the function $H(x)$ with properties mentioned above is a first integral for (4.1), and the domain G is like in Fig. 5. Then we have for $x \in G$ such that $H(N) \leq H(x) \leq H(M)$*

$$\lim_{\epsilon \downarrow 0} u_\epsilon(x) = \vartheta(H(x)),$$

where $\vartheta(y)$ is the solution of the problem:

$$\begin{aligned} \bar{L}\vartheta(y) &= 0, \quad y \in (H(N), H(M)), \\ \vartheta(H(N)) &= \psi(N), \quad \vartheta(H(M)) = \psi(M). \end{aligned} \tag{4.7}$$

The **proof** is simple, if we take into account weak convergence of Y^ϵ to Y. and the probabilistic representations of the solutions of problems (4.4) and (4.7).

Remark 4.1 Trajectories of the dynamical system starting at $x \in G$ such that $H(x) \notin [H(N), H(M)]$ leave the domain G in a finite time, and $\lim_{\epsilon \downarrow 0} u^\epsilon(x)$ can be calculated in the same way as in the Levinson case.

Remark 4.2 If a component of the boundary of the domain G, let us say, ∂G_+ coincides with a whole trajectory of the dynamical system, then the boundary condition in problem (4.7) at the point $H(M)$ should be calculated as some averaged value of $\psi(x)$, $x \in C(H(M))$. The averaging should take into account the time spent by the trajectory X_t^ϵ for $\epsilon \ll 1$ in a given part of $C(H(M))$ and how it is probable for the trajectory to exit the domain near a given point $x \in C(H(M))$. The probability of the exit near $x \in C(H(M))$ is governed by the diffusion coefficient of the transversal component of X_t^ϵ. Therefore the boundary value at the point $H(M)$ is calculated as follows:

$$\vartheta(H(M)) = \frac{1}{\bar{T}(H(M))} \int_{C(H(M))} \psi(x) \frac{\sum_{i,j=1}^2 a^{ij}(x)\, n_i(x)\, n_j(x)\, dl}{|b(x)|},$$

$$\bar{T}(H(M)) = \int_{C(H(M))} b(x)|^{-1} \sum_{i,j=1}^2 a^{ij}(x)\, n_i(x)\, n_j(x)\, dl,$$

where $n_i(x), i = 1, 2$, are the coordinates of the normal to $C(H(M))$ at $x \in C(H(M))$.

In the problem under consideration the fast motion is close to a deterministic dynamical system and the slow motion is a diffusion process. One can consider problems where the fast motion is a stochastic process, and the slow motion is close to a deterministic one or also random. We restrict ourselves to an example.

In the strip $\{x \in \mathbb{R}^2 : -\infty < x^1 < \infty, |x^2| \leq 1\} = M$ consider the diffusion process with normal reflection on the boundary $x^2 = \pm 1$ governed inside M by the operator

$$L^\epsilon = \frac{1}{2\epsilon} \frac{\partial^2}{(\partial x^2)^2} + b(x_1, x_2) \frac{\partial}{\partial x^1}, \quad 0 < \epsilon \ll 1$$

The fast motion here is the one-dimensional process in $-1 \leq x^2 \leq 1$ with reflection in the ends of the interval governed inside by the operator $\frac{1}{2\epsilon} \frac{\partial^2}{(\partial x^2)^2}$. The Lebesgue measure on $[-1, 1]$ is the invariant measure for this process.

Using the same arguments as before, one can prove that the slow component, that is the motion along x^1, converges weakly to the deterministic dynamical system in axis x^1

$$\dot{X}_t^1 = \bar{b}(X_t^1), \quad \bar{b}(x^1) = \frac{1}{2} \int_{-1}^1 b(x^1, x^2) \, dx^2 \tag{4.8}$$

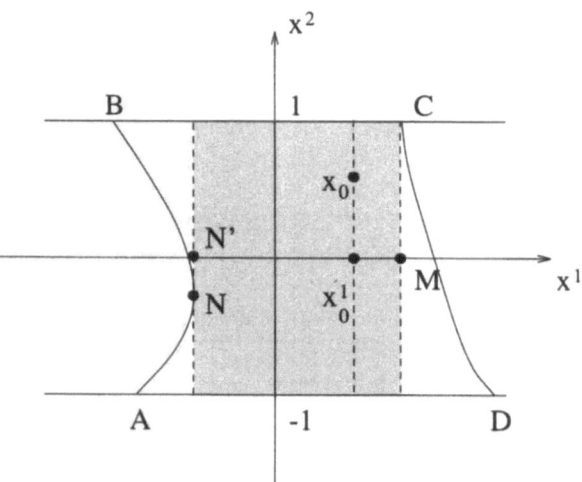

Figure 6

Such results allow to consider some PDE problems. Let G be a domain in $\mathbb{R}^2$ shown in Fig. 6; here BC and AD are parallel to axis x^1, AB and CD are smooth curves. Consider, for instance, the following small parameter problem with mixed boundary conditions:

$$L^\epsilon u^\epsilon(x) = 0, \quad x \in G,$$
$$\left.\frac{\partial u^\epsilon(x)}{\partial x^2}\right|_{x^2=\pm 1} = 0, \tag{4.9}$$
$$u^\epsilon(x)\big|_{x \in (AB \text{ or } CD)} = \psi(x),$$

where $\psi(x)$ is a continuous function.

The most right point of the curve AB is denoted by N, and let N' be the projection of N on axis X^1. We assume that the point C is the most left point of CD.

Using the above mentioned convergence of the slow motion of the process in M to the dynamical system (4.8) and the probabilistic representations of the solutions, it is easy to prove, that if the trajectory of equation (4.8), starting at x_0^1, leaves the interval $(N'M)$ (see Fig. 6) through N', then $u^\epsilon(x_0) \to \psi(N)$. If the trajectory leaves $[N'M]$ through M, then $u^\epsilon \to \psi(C)$ as $\epsilon \downarrow 0$. For points $x = (x_0^1, x_0^2)$ outside the shadowed rectangle in Fig. 6 $\lim_{\epsilon \downarrow 0} u^\epsilon(x)$ is equal to the value of $\psi(x)$ at the intersection point of the strait line $x^1 = x_0^1$ with the curves AB and CD nearest to (x_0^1, x_0^2).

If the function $\bar{b}(x^1)$ has inside $[N'M]$ zeros where it changes sign from plus to minus then the hitting time of the part of the boundary where the Dirichlet conditions are given goes to ∞ as $\epsilon \downarrow 0$. This case should be considered from the large deviation point of view (see §3). We will formulate the result in this case assuming that $\bar{b}(0) = 0$, $\bar{b}(x^1) > 0$ for $x^1 < 0$ and $\bar{b}(x^1) < 0$ for $x^1 > 0$.

Consider the eigenvalue problem depending on parameters $\alpha \in \mathbb{R}^1$ and $x^1 \in [-1,1]$:

$$\frac{1}{2}\phi''(x^2) + \alpha b(x^1, x^2)\phi(x^2) = \lambda \cdot \phi(x^2), \quad |x^2| < 1, \quad \phi'(\pm 1) = 0 \tag{4.10}$$

Let $\lambda = \lambda(x^1, \alpha)$ be the eigenvalue corresponding to the non-negative eigenfunction of problem (4.10). Such a $\lambda(x^1, \alpha)$ exists and is unique; it smoothly depends on x^1 and α if $b(x^1, x^2)$ is smooth (we now assume this), and $\lambda(x^1, \alpha)$ is convex as function of $\alpha \in \mathbb{R}^1$. Define $L(x^1, \beta)$ as Legendre transformation of $\lambda(x^1, \alpha)$ in α:

$$L(x^1, \beta) = \max_{\alpha \in \mathbb{R}^1} [\alpha\beta - \lambda(x^1, \alpha)], \quad \beta \in \mathbb{R}^1$$

Define a functional $S_{0T}(\phi)$, $\phi \in C_{0T}(\mathbb{R}^1)$ (action functional) as follows

$$S_{0T}(\phi) = \begin{cases} \int_0^T L(\phi_s, \dot{\phi}_s)\, ds & \phi \in C_{0T}(\mathbb{R}^1), \ \phi \text{ is absolutely continuous,} \\ +\infty & \text{for the rest of } C_{0T}(\mathbb{R}^1) \end{cases}$$

Let

$$V_- = \inf\{S_{0T}(\phi) : \phi_0 = 0, \ \phi_T = N', \ T > 0\}$$
$$V_+ = \inf\{S_{0T}(\phi) : \phi_0 = 0, \ \phi_T = M, \ T > 0\}$$

Then one can prove (See [F6]) that for $x = (x^1, x^2), N' < x^1 < M$,

$$\lim_{\epsilon \downarrow 0} u^\epsilon(x) = \psi(N) \quad \text{if} \quad V_- < V_+$$

$$\lim_{\epsilon \downarrow 0} u^\epsilon(x) = \psi(C) \quad \text{if} \quad V_- > V_+$$

One can consider the case of many zeros of the function $\bar{b}(x^1)$ as well.

Note that the x^2-component in problem (4.9) plays a similar part as the fast jumping component in problem (3.19).

Let us now consider shortly the averaging principle for the linear RDE-systems. We will follow the paper [EF3].

Consider again Dirichlet problem (2.10) for the system of PDE's. Assume that each of the fields $b_k(x) = (b_k^1(x), \ldots, b_k^r(x)), k = 1, \ldots, n$, has a smooth function $H(x)$ as a first integral :

$$\nabla H(x) \cdot b_k(x) = 0, \quad x \in \mathbf{R}^r, \quad k = 1, \ldots, n$$

Assume that $H(x) > 0$ for $x \neq 0$, $H(0) = 0$, and let the origin 0 be the only critical point of $H(x)$ in $\mathbf{R}^r$. Moreover, let all the level sets $C(y) = \{x \in \mathbf{R}^r : H(x) = y\}, y > 0$ be smooth manifolds homeomorphic to the sphere in $\mathbf{R}^r$.

Assume for brevity that the boundary of the domain G, where the Dirichlet problem (2.10) is considered, consists of two components ∂G_- and ∂G_+, $\partial G_- \cap \partial G_+ = \emptyset$, and let each of them be a smooth manifold of C^3-class homeomorphic to the sphere. Let

$$\min_{x \in \partial G_+} H(x) = H(x^+), \quad \max_{x \in \partial G_-} H(x) = H(x^-)$$

and let x^+ (x^-) be the unique point of absolute minimum (maximum) of $H(x)$ on ∂G_+ (∂G_-).

The fast motion now is a stochastic process (X_t^0, ν_t^0) in the phase space $C(y) \times \{1, \ldots, n\}$ (random evolution process) with y defined by the slow component. The fast motion is the Markov process governed by the operator

$$\mathcal{L}_y \vartheta(x, k) = b_k(x) \cdot \nabla_x \vartheta(x, k) + \sum_{j=1}^n c_{kj}(x)(\vartheta_j - \vartheta_k),$$
$$x \in C(y), \quad k = \{1, \ldots, n\}$$

Since $b_k(x) \cdot \nabla H(x) \equiv 0$, this process never leaves the level set, where it starts.

One can prove that the following conditions are sufficient for existence and uniqueness of an invariant measure of the process (X_t^0, ν_t^0) on $C(y) \times \{1, \ldots, n\}$: $c_{ij}(x) > 0$ for $i, j \in \{1, \ldots, n\}$, $x \in C(y)$, and the convex envelope of the vectors $b_k(x), k = \{1, \ldots n\}$, considered in $\mathbf{R}^{r-1}$ should contain the origin inside itself for any $x \in C(y)$. (Note, that all $b_k(x)$ are orthogonal to $\nabla H(x)$, so that the convex envelope is situated in the $(r-1)$-dimensional space orthogonal to $\nabla H(x)$). The invariant measure will always be finite, so that it can be normalized. Let $\mu_y(dx, i)$ be the normalized invariant measure on $C(y) \times \{1, \ldots, n\}$ for the process (X_t^0, ν_t^0).

Then, using the same arguments as earlier in this section, one can prove that the slow component $Y_t^\epsilon = H(X_t^\epsilon)$ converges weakly to the averaged process Y_t, which is governed by the operator

$$\bar{L} = \frac{1}{2}\bar{\sigma}^2 \frac{d^2}{dy^2} + \bar{b}(y)\frac{d}{dy}; \quad \bar{b}(y) = \sum_{k=1}^n \int_{C(y)} L_k H(x) \mu_y(dx, k),$$

$$\bar{\sigma}^2(y) = \sum_{k=1}^n \int_{C(y)} \sum a_k^{ij}(x) \frac{\partial H(x)}{\partial x^i} \frac{\partial H(x)}{\partial x^j} \mu_y(dx, k).$$

Denote by $\tau^\epsilon = \min\{t : H(X_t^\epsilon) \notin (H(x^-), H(x^+))\}$ (we assume that ∂G_- is situated inside the set bounded by ∂G^+). Using the weak convergence of $Y_t^\epsilon = H(X_t^\epsilon)$ to Y_t and the representation $u_k^\epsilon(x) = E_{x,k} \psi_{\nu_{\tau_\epsilon}^\epsilon}(X_{\tau^\epsilon}^\epsilon)$, one can easily prove that, if $\psi_k(x^+) = \psi^+$ and $\psi_k(x^-) = \psi^-$ are independent of k, then

$$\lim_{\epsilon \downarrow 0} u_k^\epsilon(x) = \vartheta(H(x)), \quad k = 1, \ldots, n; \quad H(x) \in [H(x^-), H(x^+)]$$

where the function $\vartheta(y)$ is the solution of the problem

$$\bar{L}\vartheta(y) = 0, \quad H(x^-) < y < H(x^+), \quad \vartheta(H(x^-)) = \psi^-, \quad \vartheta(H(x^+)) = \psi^+.$$

But if $\psi_k(x^+)$ or $\psi_k(x^-)$ are different for different k, the situation is more complicated: to calculate the limit one needs the limiting distribution of the position of $\nu_{\tau_\epsilon}^\epsilon$. This problem is still open. Some partial results one can find in [EF3].

In this section we considered the averaging principle for problems, where the fast motion has one invariant measure. The situation becomes much more complicated if the fast motion has many invariant measures. This case is considered in the next section.

In conclusion of this section I will mention one more averaging problem which is a generalization of results discussed above. Consider a differential equation

$$\dot{X}_t^\epsilon = b(X_t^\epsilon, \zeta_{t/\epsilon}), \quad X_t^\epsilon = x \in \mathbf{R}^r; \tag{4.11}$$

here $b(x, y)$ is a Lipschitz continuous function, $0 < \epsilon \ll 1$. Assume that uniformly in $x \in \mathbf{R}^r$

$$\lim_{T \to \infty} \frac{1}{T} \int_0^T b(x, \zeta_s)\, ds = b(x). \tag{4.12}$$

Then one can prove that, for any $0 < T < \infty$,

$$\max_{0 \le t \le T} |X_t^\epsilon - X_t| \longrightarrow 0 \quad \text{as } \epsilon \downarrow 0, \tag{4.13}$$

where X_t is the solution of the averaged equation

$$\dot{X}_t = b(X_t), \quad X_0 = x \in \mathrm{R}^r \tag{4.14}$$

Assumption (4.12) holds, for instance, if ζ_s is a periodic function. It is the classical averaging principle for ordinary differential equations [BM]. But (4.12) holds also when ζ_s is a stationary stochastic process with some mild mixing properties. For example, let $b(x) = E\, b(x, \zeta_t)$ and assume that

$$E[b^i(x, \zeta_s) - b^i(x)][b^j(x, \zeta_t) - b^j(x)] \longrightarrow 0$$

as $t - s \rightarrow \infty$ uniformly in x. Then (4.12) also holds, but the limit in (4.12) one should understand, say, in probability. One should consider the convergence in (4.13) also in probability. But there is an important difference between these two cases, periodic ζ_s and stationary process ζ_s with some mixing properties. To explain this, let us assume that a point O is an asymptotically stable equilibrium of the field $b(x)$ and the initial point x is attracted to O. Then for the periodic ζ_t and $\epsilon > 0$ small enough X_t^ϵ will be also attracted to a small neighborhood of O and never will leave this neighborhood. If ζ_t is a stationary process with some mixing properties, satisfying non-degeneration conditions, then X_t^ϵ also will be, first, attracted to O. But then, sooner or later, the trajectory X_t^ϵ will leave the neighborhood of O and can, for example, go to another attractor of the dynamical system (4.14). Therefore in the stochastic case the averaging principle is insufficient. One should supplement it with a description of the deviations from system (4.14).

It was proved in [Kh2] (see also [FW1]) that, under certain assumptions concerning the mixing properties of ζ_t, the process

$$\eta_t^\epsilon = \frac{X_t^\epsilon - X_t}{\sqrt{\epsilon}}$$

converges weakly in C_{0T}, $T < \infty$ as $\epsilon \downarrow 0$ to a Markov Gaussian process η_t. But this result does not help to consider the behaviour of X_t^ϵ on large time intervals or the deviations X_t^ϵ from X_t of order 1 as $\epsilon \downarrow 0$.

The theory describing the transitions of X_t^ϵ between attractors of system (4.14) and large deviations X_t^ϵ from X_t was developed in [F2](see also [FW1]).

Let, for example, ζ_t be a continuous time Markov chain with N states and intensities of transitions $c_{ij} > 0, i \neq j, c_{ii} = -\sum_{\substack{j=1 \\ j \neq i}}^{N} c_{ij}$. This example is closely connected with the problems considered in the end of the previous section. For $\alpha \in \mathrm{R}^r$, consider the matrix

$$(c_{ij} + \delta_{ij} (\alpha \cdot b(x, i))) = \hat{C}(x, \alpha),$$

where $\delta_{ij} = 1$ if $i = j$ and $\delta_{ij} = 0$ otherwise. Let $\lambda = \lambda(x, \alpha)$ be the eigenvalue of $\hat{C}(x, \alpha)$ corresponding to the eigenvector with positive elements. Such a $\lambda(x, \alpha)$

exists, differentiable and convex in α. Denote by $L(x, \beta)$ the Legendre transform of $\lambda(x, \alpha)$ in α:

$$L(x, \beta) = \sup_{\alpha \in \mathbb{R}^r} ((\alpha \cdot \beta) - \lambda(x, \alpha)), \quad \beta \in \mathbb{R}^r$$

Let

$$S_{0T}(\phi) = \begin{cases} \int_0^T L(\phi_s, \dot{\phi}_s\, ds, & \phi \in C_{0T}(\mathbb{R}^r) \quad \text{absolutely continuous,} \\ +\infty, & \text{otherwise.} \end{cases}$$

Then $\epsilon^{-1} S_{0T}(\phi)$ is the action functional for the family X_t^ϵ in C_{0T} as $\epsilon \downarrow 0$. Let

$$V(x, y) = \inf\{S_{0T}(\phi) : \phi \in C_{0T}, \phi_0 = x, \phi_T = y, T > 0\}.$$

Now, without additional assumptions, the function $V(x, y)$ can be equal to $+\infty$ for some x and y. Therefore assume that, for any $x \in \mathbb{R}^r$, the convex envelope of the vectors $\{b(x, 1), \ldots, b(x, N)\}$ contains the origin as an interior point. Then the function $V(x, y)$ is finite for any $x, y \in \mathbb{R}^r$ and even continuous.

We say that $x \sim y$ if $V(x, y) = V(y, x) = 0$. Let us denote by $\bar{\mathbb{R}}^r$ the space $\mathbb{R}^r$ with identified equivalent points. If not 'too many' points are identified, then the function $V(x, y)$ allows to answer many questions concerning long time behavior of X_t^ϵ: to describe the sequence of transitions between the attractors of the limiting dynamical system, times of these transitions, sublimiting distributions, the hierachy of cycles [F2], [FW1]. It is exactly the description which should supplement the averaging principle in the case of stochastic ζ_s. But there are some important classes of systems for which 'too many' points should be identified.

For example, let $r = 2$ and let a smooth function $H(x)$ be a first integral for system (4.14). Assume that $H(x)$ has one critical point – a minimum at the origin 0 – and that the level sets are bounded. Moreover, assume that $b(x)$ has no rest points beside $x = 0$. Then, under some mild additional assumptions concerning $b(x, \zeta_t)$, all points of the phase space $\mathbb{R}^2$ will be equivalent.

A version of the averaging principle can be used in this case. The point is that there are three time scales in our problem: the fastest motion is the motion of $\zeta_{t/\epsilon}$, the motion of X_t has the speed of order 1 as $\epsilon \downarrow 0$, and the change in time of $H(X_t^\epsilon)$ has a rate of order $o(1)$ as $\epsilon \downarrow 0$. Let us change the time to make the speed of the slowest motion of order 1 as $\epsilon \downarrow 0 : Z_t^\epsilon = X_{t/\epsilon}^\epsilon$. Then Z_t^ϵ satisfies the equation

$$\dot{Z}_t^\epsilon = \frac{1}{\epsilon} b(Z_t^\epsilon, \zeta_{t/\epsilon^2})$$

Now, before $H(Z_t^\epsilon) = Y_t^\epsilon$ changes a little, the system has enough time for averaging not only with respect to ζ_t but in the periodic motion along the level sets $C(y) = \{x : H(x) = y\}$ as well. This double averaging, combined with a central-limit-theorem type result, leads to the weak convergence of $H(Z_t^\epsilon) = Y_t^\epsilon$, $0 \le t \le T$, to a diffusion process in $\mathbb{R}^1$ (provided some assumptions on mixing of ζ_t are made).

To describe the limiting process we need some notations. Let

$$
\begin{aligned}
g(x, z) &= b(x, z) - b(x) , \\
F(x, z) &= \nabla H(x) \cdot g(x, z) , \\
D(x, z) &= EF(x, \zeta_s)F(x, \zeta_0) , \\
Q(x, s) &= E\left(\nabla F(x, \zeta_s) \cdot g(x, \zeta_0)\right) .
\end{aligned}
$$

Here ∇ denotes the gradient in x. Set

$$
D(x) = 2 \int_0^\infty D(x, s)ds, \quad Q(x) = 2 \int_0^\infty Q(x, s)ds.
$$

Under some assumptions on growth of the components of $b(x, z)$, the function $H(x)$ and their derivatives, as well as assumptions concerning the mixing properties of the process ζ_s, one can check that $D(x)$ and $Q(x)$ are finite Lipschitz continuous functions. In a sense, the functions $D(x)$ and $Q(x)$ characterize the displacement transversal to the level sets of the first integral after the averaging in the fastest motion (process $\zeta_{\frac{t}{\epsilon^2}}$). Now we can introduce the averaging in the periodic motion of system (4.14). This motion now has speed of order $\frac{1}{\epsilon}$. Let

$$
\begin{aligned}
\sigma^2(y) &= \frac{1}{T(y)} \oint_{C(y)} \frac{D(x)dl}{|b(x)|} , \\
B(y) &= \frac{1}{T(y)} \oint_{C(y)} \frac{Q(x)dl}{|b(x)|} , \\
T(y) &= \oint_{C(y)} \frac{dl}{|b(x)|} ,
\end{aligned}
\qquad (4.15)
$$

where dl is the length element on $C(y) = \{x \in \mathrm{R}^2 : H(x) = y\}$. Note that $T(y)$ is the period of the rotation along the curve $C(y)$. Then the process $Y_t^\epsilon = H(Z_t^\epsilon)$ converges weakly in C_{OT} to the diffusion process Y_t,

$$
dY_t = \sigma(Y_t)dW_t + B(Y_t)dt;
$$

here W_t is an one-dimensional Wiener process. This result is proved in [BF].

I am going to describe now a similar result for a multidimesional dynamical system (4.11) with $l > 1$ conservation laws. Let the corresponding averaged system (4.14) have l first integrals $H_1(x), \ldots, H_l(x)$. Assume that all the functions $H_k(x)$ are smooth enough and that the vectors $\nabla H_l(x), \ldots, \nabla H_l(x)$ are linearly independent. Assume, that the set $C(y) = \{x \in \mathrm{R}^r : H_1(x) = y_1, \ldots, H_l(x) = y_l\}$ is compact and connected for any $y = (y_1, \ldots, y_l)$ ($C(y)$ can be empty). Let a measure μ_y on $C(y)$ exist such that $\mu_y\left(C(y)\right) = 1$ and for any continous function $f(x)$ on $C(y)$

$$\lim_{T \to \infty} \frac{1}{T} \int_0^T f(X_s)ds = \int_{C(y)} f(x)\mu_y(dx) \qquad (4.16)$$

uniformly in the inital point $X_0 = x \in C(y)$. Condition (4.16) replaces the periodicity condition. Then, under certain assumptions concerning the mixing rate for the process ζ_t, one can prove that the processes $H(X_{t/\epsilon}^\epsilon) = (H_1(X_{t/\epsilon}^\epsilon), \ldots, H_l(X_{t/\epsilon}^\epsilon))$ converge weakly in the space of continuous functions on $[0, T]$ to a l-dimensional diffusion process Y_t. The diffusion and drift coefficients for the limiting process can be expressed in a form similar to (4.16) (see [BF]). But condition (4.16) turns out to be too restrictive if the dimension of the sets $C(y)$ is bigger then 1. Probably, the convergence can be proved if (4.16) is replaced by a weaker assumption, for example, if (4.16) is fulfilled for almost all y and for almost all initial points $X_0 = x \in C(y)$. But this problem is still open. Problem (4.11) was studied in [S] and [Kh1] under the assumption that $Eb(x, \zeta_s) \overset{x}{\equiv} 0$. It means that each coordinate $x_k = H_k(x)$, $k = 1, \ldots, r$, is a first integral. In this case, $C(y)$ for any $y \in \mathbb{R}^r$ the set $C(y)$ consists of one point and condition (4.16) is fulfilled automatically.

If the set $C(y)$ consists of more than one connected component or, in general, if the averaged system (4.14) has on $C(y)$ more than one 'smooth' invariant measure, the processes $Y_t^\epsilon = H(X_{t/\epsilon}^\epsilon)$ will not converge to a Markov process. We consider such questions in the next section.

5 Averaging Principle: Continuation

Consider a Hamiltonian dynamical system in the plane R^2

$$\dot{X}_t = \bar{\nabla} H(X_t), \ X_0 = x \in R^2, \ \bar{\nabla} H(x) = \left(\frac{\partial H(x)}{\partial x^2}, -\frac{\partial H(x)}{\partial x^1} \right). \tag{5.1}$$

We assume that $H(x) \to \infty$ as $|x| \to \infty$, that $H(x)$ has a finite number of critical points and $\min_{x \in R^2} H(x) = 0$. Then all the level sets

$$C(y) = \{ x \in R^2 : H(x) = y \}, \ y \geq 0$$

are compact. Each $C(y)$ consists of a finite number $n(y)$ of connected components $C_i(y)$:

$$C(y) = \bigcup_{i=1}^{n(y)} C_i(y)$$

If y is not a critical value of $H(x)$ then each component $C_i(y)$ is a periodic trajectory of system (5.1).

For brevity let $H(x)$ be a generic function: all its critical points are non-degenerate and each level set $C(y)$ contains at most 1 critical point. If $C(y)$ contains a critical point x_0, then $C(y)$ can contain two more tajectories having x_0 as their limit as $T \to \pm\infty$ besides the trajectory $X(t) \equiv x_0$. One can check that all points of the phase space for such a system are equivalent (from the large deviation point of view) with respect to the white noise perturbations. The simplest example is given by the harmonic-oscillator-type Hamiltonian, where $H(x)$ has only one critical point, let us say, at the origin (Fig. 7a).

The corresponding phase picture is given in Fig. 7b: each level set $C(y)$ consists of one periodic trajectory. A unique normalized invariant measure of the system concentrated on $C(y)$ has the density

$$M_y(x) = \frac{1}{T(y)} |\nabla H(x)|^{-1}, \quad y > 0, \ x \in C(y) \tag{5.2}$$

where $T(y)$ is the period of the trajectory on the level

$$y : T(y) = \int_{C(y)} |\nabla H(x)|^{-1} \, dl,$$

dl is the length element on $C(y)$.

Let us consider now the case when $H(x)$ has more then one critical point (Fig. 8a).

The set of trajectories in that case consists of several families of periodic orbits devided by the separatrices. For example, in Fig. 8b there are five families:

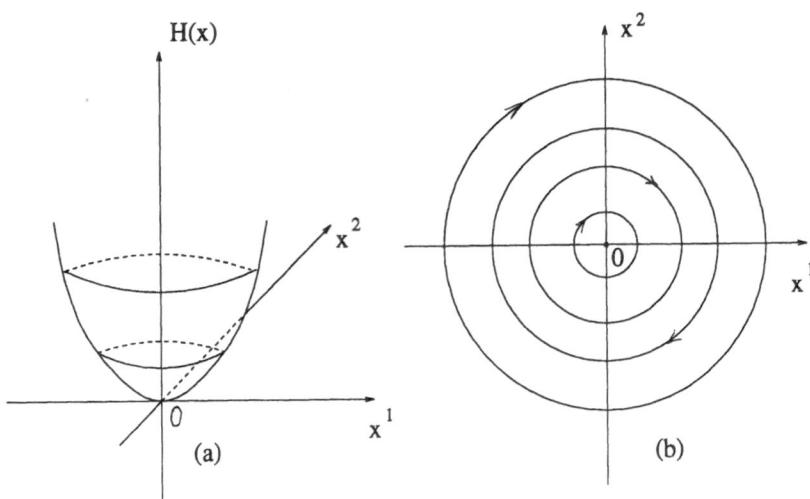

Figure 7

rotations around O_1, around O_3 and around O_5, periodic orbits containing the three points O_1, O_2, O_3, and periodic orbits around all five critical points. These families are separated by two ∞-shaped curves: γ_1 with crossing point at O_2, and γ_2 with crossing point at O_4 (Fig. 8d).

An important feature of the system with Hamiltonian having many critical points is the appearence of a new first integral independent of $H(x)$. This integral $\hat{H}(x)$ is the number of the family of the periodic orbits:

$$\hat{H}(x) = \begin{cases} 1 & \text{, if } X^x \text{ is surrounded by the left loop of } \gamma_1 \\ 2 & \text{, if } X^x \text{ is surrounded by the right loop of } \gamma_1 \\ 3 & \text{, if } X^x \text{ is surrounded by the right loop of } \gamma_2 \\ 4 & \text{, if } X^x \text{ is situated inside the left loop of } \gamma_2 \\ & \quad \text{and } \gamma_1 \text{ is inside the loop } X^x \\ 5 & \text{, if } \gamma_2 \text{ is inside the loop } X^x \end{cases}$$

Here X^x is the whole trajectory starting at $x \in \mathrm{R}^2$.

It is clear that $\hat{H}(X_t^x)$ does not change with time. Of course, any function of $H(x), \hat{H}(x)$ is also a first integral.

Consider the white noise perturbations of the system (5.1):

$$\dot{\tilde{X}}_t^\epsilon = \bar{\nabla} H(\bar{X}_t^\epsilon) + \epsilon \dot{W}_t$$

As we did before, rescale the time: $X_t^\epsilon = \tilde{X}_{t/\epsilon^2}^\epsilon$. Then we have for X_t^ϵ the following equation

$$\dot{X}_t^\epsilon = \frac{1}{\epsilon^2} \bar{\nabla} H(X_t^\epsilon) + \dot{W}_t. \tag{5.3}$$

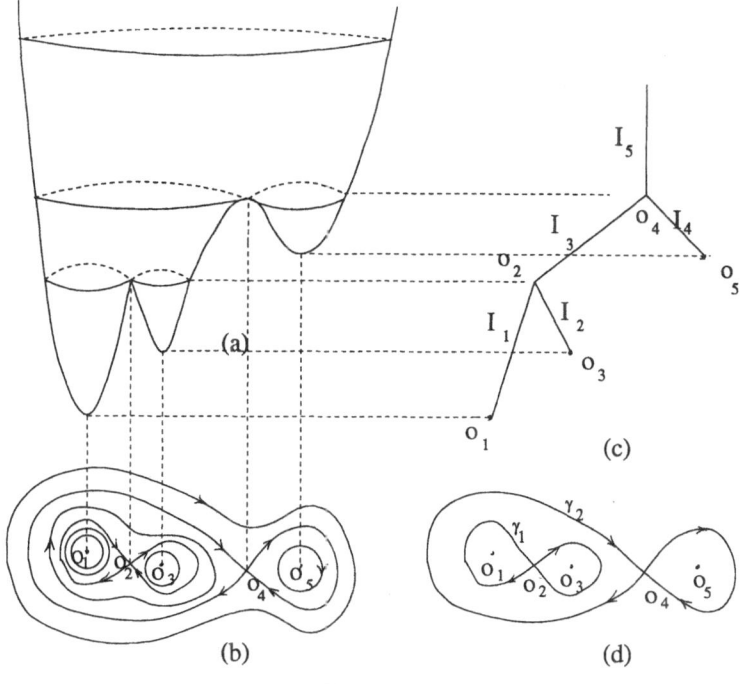

Figure 8

We have seen already that in the case of one critical point the slow component $H(X_t^\epsilon) = Y_t^\epsilon$ weakly converges to the one-dimensional diffusion process Y_t in the half-line governed by the operator

$$\bar{L} = \frac{\bar{\sigma}^2(y)}{2} \frac{d^2}{dy^2} + \bar{b}(y) \frac{d}{dy},$$ (5.4)

where $\bar{\sigma}^2(y)$ and $\bar{b}(y)$ are calculated by averaging with respect to the invariant density (5.2) on the level set $C(y)$:

$$\bar{b}(y) = \frac{1}{2T(y)} \int_{C(y)} \frac{\Delta H(x)\, dl}{|\nabla H(x)|}, \quad \bar{\sigma}^2(y) = \frac{1}{T(y)} \int_{C(y)} |\nabla H(x)|\, dl.$$

Note that, in the case of one critical point, $C(y)$ consists of just one connected component. The process Y_t moves in $\mathrm{R}^+ = \{x \in \mathrm{R}^1, x \geq 0\}$; the point $x = 0$ is inaccessible from the inside of R^+.

Suppose now that $H(x)$ has several critical points. In this case $C(y)$ consists, at least for some y, of several connected components $C_1(y), \ldots, C_{n(y)}(y)$. An invariant measure is concentrated on each of these components and one should average the coefficients not over the whole level set $C(y)$ but just over the component containing the trajectory. The behavior of the process $H(X_t^\epsilon)$ before a time t_0 can help us to identify the connected component containing the process at time t_0. Therefore one cannot expect that $H(X_t^\epsilon)$ converges in this case to a Markov process: the behavior of the process after time t_0, given the position at t_0, depends on the behavior of the process before t_0.

To have a Markov process in the limit, one has to extend the phase space. One should include in the phase state, in addition to $H(X_t^\epsilon)$, the value of the additional first integral $\hat{H}(x)$. Then at time t_0 we will know not only the level set, but the connected component of this level set. The dynamical system has some mixing properties on each connected component. This allows us to hope that the process $(H(X_t^\epsilon), \hat{H}(X_t^\epsilon))$ will converge to a Markov process as $\epsilon \downarrow 0$.

The set of connected components of level sets of the Hamiltonian $H(x)$, provided with the natural topology, is homeomorphic to a graph Γ. For example, in Fig. 8 each minimum point of $H(x)$ corresponds to an exterior vertex O_1, O_3 or O_5 of the graph. The saddle points together with ∞-shaped curves correspond to the interior vertices O_2 and O_4. The points of the open edges $I_1, \ldots, I_5$ correspond to the periodic orbits. Say, I_1 counts all orbits around O_1 up to the energy level $H(O_2)$, and I_2 corresponds to the rotations around O_3 up to the energy $H(O_2)$. The points of I_3 correspond to the orbits in the region where $\hat{H}(x) = 4$.

For systems in R^2 (or, in general, in R^r) the corresponding graph has the structure of a tree. If some critical points of $H(x)$ degenerates, more than 3 edges can meet at the vertex of the graph corresponding to that point. For Hamiltonians on manifolds with a more complicated topological structure, say on a torus, the graph can have loops.

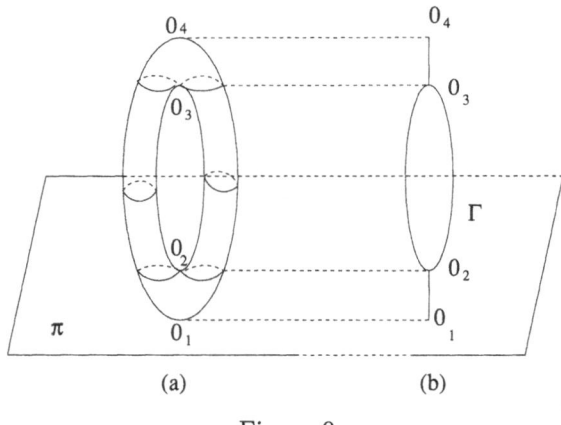

(a)   (b)

Figure 9

Consider, for example, a Hamiltonian dynamical system on the two-dimensional torus T^2. Let the torus be invariant with respect to the rotation around a straight line L parallel to a plane π (Fig. 9a), and let $H(x)$ be equal to the distance of a point $x \in T^2$ from the plane π. The level sets of such a function $H(x)$ are drawn in Fig 9a. Then, the family of connected components of these level sets is homeomorphic to the graph Γ shown in Fig 9b. The graph has a loop in this case.

Denote by Y the mapping of the set of connected components of the level sets of $H(x)$ to the graph $\Gamma : Y(C_i(y))$ is the point of I_i corresponding to $C_i(y)$. One can consider the value of $H(x)$ as a coordinate in I_i, so that $Y(C_i(y))$ is the pair (y, i), which characterizes a point of the graph.

One can extend the mapping Y to the mapping $R^2 \to \Gamma : Y(x)$, $x \in R^2$, is defined as the point of Γ corresponding to the connected component $C_{i(x)}(H(x))$ of the level set $C(H(x))$ containing the point $x : Y(x) = (H(x), i(x))$.

Consider the family of stochastic processes on the graph Γ

$$Y_t^\epsilon = Y(X_t^\epsilon), \ t \geq 0.$$

It turns out that the process Y_t^ϵ, $0 \leq t \leq T$, converges weakly in the space of continuous functions on $[0, T]$ with values in the graph Γ to a continuous Markov process on Γ as $\epsilon \downarrow 0$.

It is worth mentioning that there exists a number of "classical" asymptotic problems where the limiting process has values in a graph. One can find some of such problems in [FW2]. We will mention some of such problems later.

It is important for all of these asymptotic problems to have a description of the class of continuous Markov processes on graphs. I will give such description following [FW2].

Consider a connected graph Γ consisting of the edges $I_1, \ldots, I_n$ and vertices $O_1, \ldots, O_m$. We will write $I_i \sim O_k$, if O_k is the end of I_i. Let $L_1, \ldots, L_n$ be second order elliptic differential operators

$$L_i = \frac{1}{2} \frac{\sigma_i^2(y)}{2} \frac{d^2}{dy^2} + B_i(y) \frac{d}{dy}, \ y \in I_i.$$

We assume that the coefficients are Lipschitz continuous and bounded, $\sigma_i(y) \geq \sigma > 0$. Then a diffusion process $X_t^{(i)}$ in I_i corresponds to L_i, $i = 1, \ldots, n$. The process $X_t^{(i)}$ is defined up to the first exit from I_i. How can one describe continuous Markov processes on Γ, coinciding with $X_t^{(i)}$ inside the edges? We should define the behavior of the process after reaching a vertex.

Here the situation is similar to the well known problem considered by Feller: Describe all possible continuations of a continuous Markov process in an open interval to a process on the closed interval, preserving the continuity and the Markov property. The most convenient way to describe all such continuations is to describe the domain of definition of the generator of the extended Markov process. If the process inside an interval I is governed by the operator $L =$

$\frac{1}{2}a(x)\frac{d^2}{dx^2}+b(x)\frac{d}{dx}$, $a(x) > 0$, then each possible continuation is defined by boundary conditions in the ends of the interval. For example, if the process in the closed interval has the instantanious reflection in the boundary, then the corresponding boundary conditions are $\frac{df(x)}{dx}\Big|_{x\in\partial I} = 0$. This means that the generator A of the extended process is defined for smooth $f(x)$, $x \in I$ such that $\frac{df(x)}{dx}\Big|_{x\in\partial I} = 0$ and $Af(x) = Lf(x)$ inside the interval. W. Feller described all possible boundary conditions corresponding to the Markovian continuations of the process inside I.

In our case the boundary conditions should be replaced by some gluing conditions at the vertices. Any function which is smooth in $\Gamma\backslash\{O_1, \ldots, O_m\}$ and satisfies these gluing conditions should belong to the domain of definition of the generator of the extended process.

It is proved in [FW2], that for any set of constants $\alpha_k, \beta_{kj} \geq 0, k \in \{1, \ldots, m\}$, $j \in \{i : I_i \sim O_k\}, \alpha_k + \sum_{i:I_i\sim O_k} \beta_{ki} \neq 0$, there exists a unique continuous Markov process Y_t on Γ such that its generator A is defined for continuous functions $f(y), y \in \Gamma$, satisfying the conditions:

1. $f(y)$ is twice continuously differentiable inside the edges $I_i \subset \Gamma$;
2. If $I_i \sim O_k$ then $\lim\limits_{\substack{y\to O_k \\ y\in I_i}} L_i f(y)$ exists and is independent of i; we denote this limit by $Lf(O_k)$;
3. $\alpha_k Lf(O_k) + \sum_{i:I_i\sim O_k} \beta_{ki}\frac{df}{dy^i}(O_k) = 0, \quad k = 1, \ldots, n$;

here y_i is the coordinate on I_i equal to the distance of the point of I_i from 0_k. If $f(y)$ satisfies these conditions, then $Af(y) = L_i f(y)$ for $y \in I_i$. Moreover, for any continuous Markov process on Γ governed by the operator L_i inside I_i, $i = 1, \ldots, n$, one can find constants $\alpha_k, \beta_{kj} \geq 0, \alpha_k + \sum_{i:I_i\sim O_k} \beta_{ki} \neq 0$ for $k = 1, \ldots, m$, such that the generator A of that process is defined for $f(y), y \in \Gamma$, satisfying conditions 1-3, and $Af(y) = L_i f(y)$ for $y \in I_i$.

If our graph consists of one segment this statement coincides with Feller's result. The coefficients α_k, β_{ki} characterize the behavior of the process at vertex $O_k \in \Gamma$. The probabilistic meaning of the coefficients is the following: α_k describes delay at O_k and β_{ki}, roughly speaking, shows how the particle will be distributed between the edges $I_i \sim O_k$ immediately after leaving O_k.

We assumed that the operators L_i that govern the process inside the edges are uniformly elliptic. This condition is fulfilled in a number of asymptotic problems where the limiting process is a Markov process on a graph. However, if we study the white noise perturbations of dynamical systems with conservation laws, then we must consider degenerate operators: the coefficients $\bar{\sigma}^2(y)$ defined above vanish at the vertices since $\nabla H(x) = 0$ at the critical points. Such a degeneration can make a vertex inaccessible for the limiting process. The gluing conditions for a general class of continuous Markov processes on graphs (including degenerate processes) are described in [FW2].

To describe the process Y_t on the graph Γ limiting for the family $Y_t^\epsilon = H(X_t^\epsilon)$ as $\epsilon \downarrow 0$, we should calculate the operators L_i for each edge $I_i \in \Gamma$ and the gluing conditions at the vertices. The calculation of the operators L_i, actually, is similar to the case of Hamiltonians with one critical point (see §4). The only difference is that in the general case, the level set $C(y)$ is a union of several connected components $C(y) = \bigcup_{i=1}^{n(y)} C_i(y)$. The averaging should now be carried out only over the corresponding component $C_i(y)$. Therefore the operator

$$L_i = \frac{\sigma_i^2(y)}{2} \frac{d^2}{dy^2} + B_i(y) \frac{d}{dy},$$

governing the limiting process Y_t inside I_i, has coefficients

$$
\begin{aligned}
\sigma_i^2(y) &= \frac{1}{T_i(y)} \int_{C_i(y)} |\nabla H(x)| \, dl, \\
B_i(y) &= \frac{1}{2T_i(y)} \int_{C_i(y)} \frac{\Delta H(x) \, dl}{|\nabla H(x)|}, \\
T_i(y) &= \int_{C_i(y)} |\nabla H(x)|^{-1} \, dl.
\end{aligned}
\tag{5.5}
$$

$T_i(y)$ is the period of the trajectory of dynamical system (5.1) if the energy $H = y$ and the starting point x is such that $Y(x) \in I_i$.

One can see from formulas (5.5) that the diffusion coefficients $\sigma_i^2(y)$ degenerate at the vertices O_k, $k = 1, \ldots, m$. Simple calculations show that the order of degeneration of the diffusion coefficients at the vertices corresponding to the extrema of $H(x)$ (exterior vertices) and the signs of the drift coefficients at these points are such that the exterior vertices $(O_1, O_3, O_5$ in Fig. 8) are inaccessible for the limiting process Y_t on the graph. It means that no additional conditions should be imposed at these points.

The situation is different at the interior vertices corresponding to the saddle points of the Hamiltonian $H(x)$ (O_2 and O_4 in Fig. 8). Although the diffusion is degenerate at these points, the degeneration is slow enough. All such points are accessible and gluing conditions should be imposed at these points.

To formulate the gluing condition at an interior vertex O_k, consider the ∞-shape curve γ_k corresponding to the saddle point O_k. Such curves correspond to each non-degenerate saddle point (if O_k is degenerate saddle point, the level set $C(H(O_k))$ consists of several loops attached to O_k, and the gluing conditions will be similar). In the non-degenerate case γ_k consists of two loops γ_k^1 and γ_k^2 (Fig. 10). Let the edge I_1 correspond to the orbits located inside γ_k^1, I_2 to the orbits located inside γ_k^2, and I_3 to the orbits containing γ_k inside themselves. Denote

$$\beta_{ki} = \int_{\gamma_k^i} |\nabla H(x)| \, dl, \ i = 1, 2 \qquad\qquad - \beta_{k3} = \beta_{k1} + \beta_{k2}$$

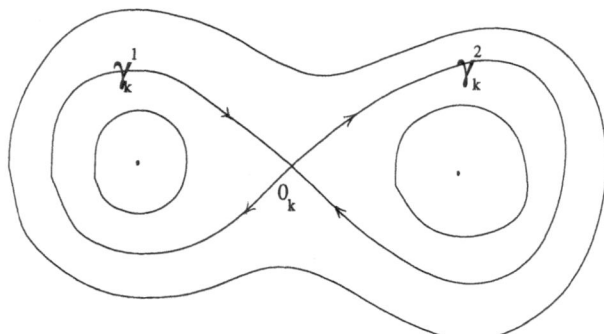

Figure 10

Then the gluing conditions at the vertex O_k have the form:

$$\beta_{k1} \frac{df}{dy_1}(O_k) + \beta_{k2} \frac{df}{dy_2}(O_k) = \beta_{k3} \frac{df}{dy_3}(O_k); \tag{5.6}$$

here y_i, $i = 1, 2, 3$ are coordinates in I_i. Note that $\alpha_k = 0$. This means that the limiting process has no delay at the vertex O_k.

The operators L_i, $i = 1, \ldots, n$, and the gluing conditions (5.6) at the interior vertices define the limiting process in a unique way. These results were proved in [FW3].

How can one prove the convergence of $Y_t^\epsilon = Y(X_t^\epsilon)$ and calculate the gluing conditions? First, one should check the tightness of the family Y_t^ϵ, $0 \le t \le T$, in the weak topology, then calculate the operators L_i, $i = 1, \ldots, n$ using the averaging procedure. The next step is to prove that the limiting process is Markovian. Now, since we have a full description of the Markov continuous processes on graphs, we should find the gluing conditions. To do this, one can use the fact that the Lebesgue measure is the invariant measure for X_t^ϵ in R^2 for any $\epsilon \ge 0$. Using this fact, one can calculate the invariant measure for the process $Y_t^\epsilon = Y(X_t^\epsilon)$ on Γ. This measure is a "projection" of Lebesgue's measure in R^2 on Γ. It is independent of ϵ. Now one should choose constants β_{ki} at each vertex O_k so that the process on Γ with given operators L_i and the gluing conditions defined by these β_{ki} have the prescribed invariant measure. A plan close to this one (but slightly different) was realized in [FW3].

The result is similar if one considers a more general class of perturbations

$$\dot{X}_t^\epsilon = \bar{\nabla} H(X_t^\epsilon) + \epsilon \sigma(X_t^\epsilon) \dot{W}_t, \quad X_0 = x \in \mathrm{R}^2,$$

where $\sigma(x)\sigma^*(x) = a(x)$ is a non-degenerate matrix with sufficiently smooth coefficients.

Thus, the evolution of the energy under the white-noise-type perturbations can be described in a proper time scale as a diffusion process on the graph corresponding to the Hamiltonian H. The limiting process has no delay at the vertices. It is not necessary, in general, that the evolutions of the first integrals have no delay at vertices of the graph; some vertices can correspond to a set in the phase space, where the non-perturbed trajectory spends a positive time. Then the limiting process has a delay at such vertices. One can find an example of this situation in the next section (see also [F11]) where the perturbations of area preserving dynamical systems on torus are considered shortly.

Consider a dynamical system (4.1) in R^2 having a smooth first integral $H(x)$. Assume that $H(x)$ has a finite number of critical points. Since $H(x)$ is a first integral for (4.1), $b(x) \cdot \nabla H(x) = 0$, and thus

$$b(x) = \beta(x)\bar{\nabla}H(x), \qquad (5.7)$$

where $\beta(x)$ is a scalar. If $\beta(x) \equiv 1$ we have the Hamiltonian system considered before. If $\beta(x) \not\equiv 1$, but does not change its sign, the situation is similar to the Hamiltonian case. But if $\beta(x)$ changes its sign, we face a more complicated problem: even if $H(x)$ is such as in §4 (i.e. has one minimum point and the level sets are topological circles) the slow evolution process for vector field (5.7) should be considered on a graph (see [F11], [BF]). The process $H(X_t^\epsilon)$, where X_t^ϵ is defined by (4.3), does not converge, in general, to a Markov one as $\epsilon \downarrow 0$.

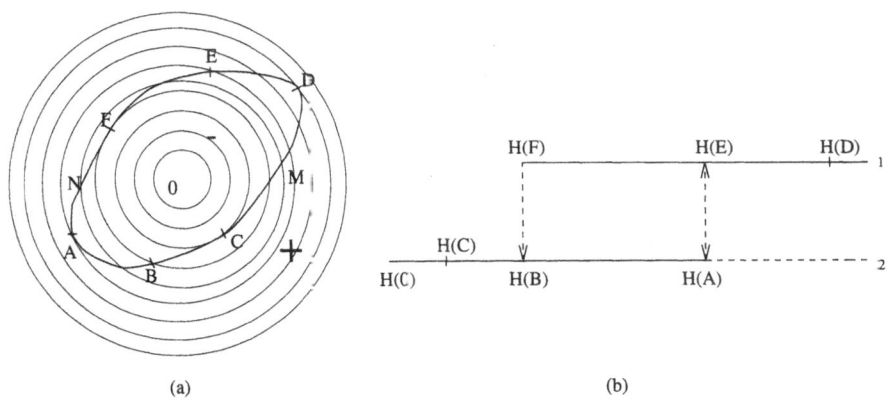

(a)   (b)

Figure 11

Let $H(x)$, $x \in R^2$ have just one minimum at the origin and $H(0) = 0$, and let the sets $C(y) = \{x \in R^2 : H(x) = y\}$, $y > 0$, be smooth curves homeomorphic to the circle. Suppose $\beta(x)$ is negative inside the loop $ABCDEFA$ (see Fig. 11a) and positive outside this loop. Here A, C, D, F are the points where the loop $ABCDEFA$ is tangent to a level set of $H(x)$; B is the point where the trajectory starting at F crosses AC, and E is the point where the trajectory starting at A crosses FD. Then the dynamical system on some of the level sets $C(y) = \{x : H(x) = y\}$ has four rest points: two stable and two unstable. This results in appearance of a new first integral independent of $H(x)$.

To define this new first integral $\nu(x)$, denote by $x_0^{(1)}(y)$ the point of intersection of $C(y) = \{x \in R^2 : H(x) = y\}$ and the arc FED of the curve $\gamma = \{x \in R^2 : \beta(x) = 0\}$. Denote the intersection point of $C(y)$ and of the arc ABC by $x_0^{(2)}(y)$. We assume that there exists at most one such $x_0^{(1)}(y)$ and at most one $x_0^{(2)}(y)$. It is clear that $x_0^{(1)}(y)$ exists for $H(F) < y < H(D)$, and $x_0^{(2)}(y)$ exists for $H(C) < y < H(A)$.
Define $\nu(x), x \in R^2$, as follows:

$$\nu(x) = \begin{cases} i, & \text{if } x_0^{(i)}(H(x)) \text{ exists and } x \text{ belongs to the domain of} \\ & \text{attraction of } x_0^{(i)}(H(x)), i = 1, 2, \\ 1, & \text{if } H(x) \geq H(D), \text{ or if x belongs to the arc CMD,} \\ 2, & \text{if } H(x) \leq H(C), \text{ or if x belongs to the arc ANF.} \end{cases}$$

It is easy to see that $\nu(x)$ is a first integral for vector field (5.7) independent of $H(x)$.

Consider now small white noise perturbations of system (4.1) with $b(x) = \beta(x)\bar{\nabla}H(x)$.

$$\dot{\tilde{X}}_t^\epsilon = \beta(\tilde{X}_t^\epsilon)\bar{\nabla}H(\tilde{X}_t^\epsilon) + \epsilon^{1/2}\dot{W}_t \qquad \tilde{X}_0^\epsilon = x \in R^2.$$

and the rescaled process $X_t^\epsilon = \tilde{X}_{t/\epsilon}^\epsilon$. Because of the existence of an independent first integral, $H(X_t^\epsilon)$ will not converge to a Markov process as $\epsilon \downarrow 0$. One has to consider an extended phase space to preserve the Markov property.

Consider the graph Γ in Fig. 11 b. The pairs (H, i), $H \geq H(O) = 0, i = 1, 2$, can be used as coordinates on Γ. The points $(H(F), 1)$ and $(H(B), 2)$ as well as $(H(E), 1)$ and $(H(A), 2)$ are identified.

Define the mapping $Y(x) : R^2 \to \Gamma$ by $Y(x) = (H(x), \nu(x))$, and let $Y_t^\epsilon = Y(X_t^\epsilon)$.

It is easy to see that, when $H(X_t^\epsilon) \notin [H(B), H(A)]$, the fast component of the process X_t^ϵ has distribution on $C(H(X_t^\epsilon))$ close to the invariant measure of the non-perturbed dynamical system on $C(H(X_t^\epsilon))$. When $H(X_t^\epsilon) \in [H(B), H(A)]]$, the fast coordinate of the X_t^ϵ is concentrated near one of the points $x_0^{(1)}(H(X_t^\epsilon))$ or $x_0^{(2)}(H(X_t^\epsilon))$, depending from which side $H(X_t^\epsilon)$ last time entered the segment $[H(B), H(A)]$. If $\epsilon \ll 1$, the process X_t^ϵ "jumps" from F to B along the deterministic trajectory. Similarly, X_t^ϵ "jumps" from A to E.

One can check, using these arguments, that the process Y_t^ϵ converges weakly as $\epsilon \downarrow 0$ on the space of continuous functions on $[0, T], T < \infty$, with the values in Γ to a diffusion process Y_t on Γ. The process Y_t is defined as follows. Let $T(y) = \int_{C(y)} [\| \beta(x)\nabla H(x) \|]^{-1} dl$, and

$$A_1(y) = \begin{cases} [T(y)]^{-1} \int_{C(y)} \frac{|\nabla H(x)| dl}{|\beta(x)|}, & y > H(D), \\ |\nabla H(x_0^{(1)}(y))|^2, & \text{if } H(F) \leq y \leq H(D); \end{cases}$$

$$B_1(y) = \begin{cases} [2T(y)]^{-1} \int_{C(y)} \frac{\Delta H(x) dl}{|\beta(x)\nabla H(x)|}, & y > H(D), \\ \frac{1}{2} |\Delta H(x_0^{(1)}(y))|, & \text{if } H(F) \leq y \leq H(D); \end{cases}$$

$$A_2(y) = \begin{cases} [T(y)]^{-1} \int_{C(y)} \frac{|\nabla H(x)| dl}{|\beta(x)|}, & y < H(C), \\ |\nabla H(x_0^{(2)}(y))|^2, & \text{if } H(C) \leq y \leq H(A); \end{cases}$$

$$B_2(y) = \begin{cases} [2T(y)]^{-1} \int_{C(y)} \frac{\Delta H(x) dl}{|\beta(x)\nabla H(x)|}, & y < H(C), \\ \frac{1}{2} |\Delta H(x_0^{(2)}(y))|, & \text{if } H(C) \leq y \leq H(A); \end{cases}$$

$$L_i = \tfrac{1}{2} A_i(y) d^2/dy^2 + B_i(y) d/dy, i = 1, 2.$$

The process Y_t on Γ is governed by the operator L_1 on the upper part of the graph Γ (i = 1), and by L_2 on the lower part (i = 2). The operator L_2 degenerates at the point $(O, 2) \in \Gamma$, so that this point is inaccessible, and no boundary condition should be imposed there.

To define $Y_t, t \geq 0$, in a unique way, we should add the gluing conditions at the points (H(F),1) identified with (H(B),2) and at (H(E),1) identified with (H(A),2). As already mentioned, the gluing conditions are described by the domain of definition of the generator for the process Y_t. Let Y_t be the process with the generator defined for functions $f(y, i)$ which are continuous on Γ, smooth on $\{(y, i) \in \Gamma : y \geq H(F), i = 1\}$ and on $\{(y, i) \in \Gamma : y \leq H(A), i = 2\}$ and satisfy $L_1 f(H(F), 1) = L_2 f(H(B), 2)$ and $L_1 f(H(E), 1) = L_2 f(H(A), 2)$. Obviously, these gluing conditions have the form described above. Together with the operators L_1 and L_2, they define the process Y_t on Γ in a unique way.

It is clear that all these results concerning convergence of stochastic processes should yield some results concerning the corresponding PDE's with a small parameter in the higher derivatives. Consider, for example, the dynamical system shown in Fig. 8b, and let G be the domain with a smooth boundary drawn in Fig. 12a. Here the boundary consists of three components $\partial G_1, \partial G_2, \partial G_3$. The trajectory $X^{(1)}$ is located inside the closure $\bar{G}$ of G with the maximal H, the trajectory $X^{(2)}$ is located in $\bar{G}$ and in the left loop of γ_1 having the minimal energy, and the trajectory $X^{(3)}$ is in the right loop of γ_2 (and inside $\bar{G}$) having the minimal H, drawn by dotted lines. Let $\psi(x)$ be a continuous function on $\partial G = \partial G_1 \cup \partial G_2 \cup \partial G_3$. Consider the Dirichlet problem

$$\frac{\epsilon^2}{2}\Delta u^\epsilon(x) + \bar{\nabla} H(x) \cdot \nabla u^\epsilon(x) = 0, \ x \in G, u^\epsilon(x)|_{\partial G} = \psi(x).$$

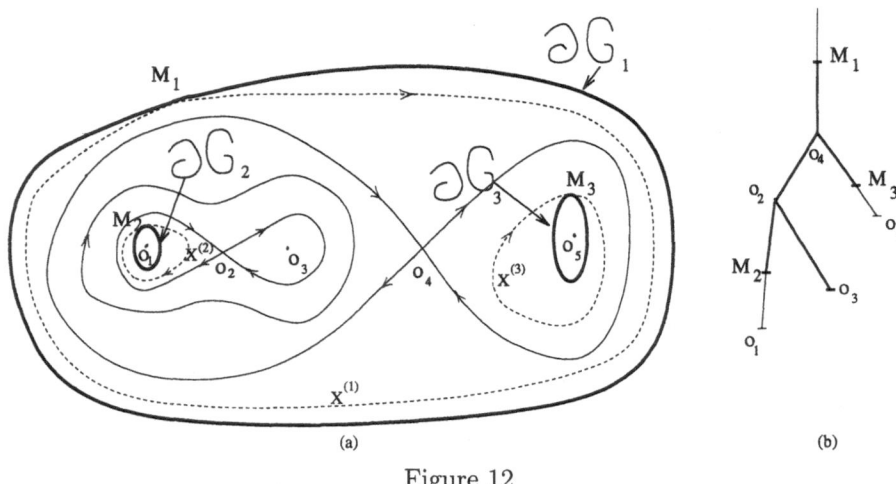

Figure 12

For points $x \in G$, located outside the domain in Fig. 12a bounded by the three dotted orbits, trajectories of the diffusion process X_t^ϵ corresponding to (5.8) starting at such x, leave G in a finite time. It is, roughly speaking, the Levinson case.

Let x be located inside the domain bounded by the dotted orbits. Consider the graph Γ corresponding to $H(x)$ (See Fig. 12b). Denote by M_i the point of ∂G where ∂G is tangent to $X^{(i)}$, $i = 1, 2, 3$; the points of the graph, corresponding to $X^{(i)}$, $i = 1, 2, 3$, are denoted by M_i. Define $\hat{\Gamma}$ as the part of the graph Γ bounded by the points M_1, M_2, M_3 and O_3 (see Fig. 12). Then from the weak convergence of $Y(X_t^\epsilon)$ to Y_t, using the probabilistic representation of the solution, one can derive that

$$\lim_{\epsilon \downarrow 0} u^\epsilon(x) = \vartheta(H(x), i(x)) = \vartheta(Y(x)),$$

where $\vartheta_i(y) = \vartheta(y, i)$ is the solution of the problem

$$L_i \vartheta_i(y) = 0, \ y \in I_i \cap \hat{\Gamma}, \ i = 1, 2, 3, 4, 5,$$

$$\vartheta(m_i) = \psi(M_i), \ i = 1, 2, 3,$$

$$\sum_{j: I_j \sim O_k} \beta_{kj} \frac{d\vartheta_j}{dy^j}(O_k) = 0, \ k = 2, 4.$$

(5.8)

Here β_{kj} are the constants defined above. Problem (5.8) has a unique solution, which, actually can be written in an explicit form.

6 Remarks and Generalizations

In this and the next section, we consider some other asymptotic problems also leading to processes on graphs and corresponding problems for PDE's.

1. Consider a Markov process (X_t, P_x) in a phase space $\mathcal{E}$. A function $H(x) : \mathcal{E} \to \mathbf{R}^1$ is called a first integral for (X_t, P_x) if

$$P_x\{H(X_t) = H(x)\} = 1, \quad x \in \mathcal{E}.$$

If (X_t, P_x) is the diffusion process in $\mathbf{R}^r$ governed by the operator

$$L = \frac{1}{2} \sum_{i,j=1}^{r} A^{ij}(x) \frac{\partial^2}{\partial x^i \partial x^j} + \sum_{i=1}^{r} B^i(x) \frac{\partial}{\partial x^i},$$

then a smooth function $H(x)$ is a first integral if and only if, for any $x \in \mathbf{R}^r$,

$$\sum_{i,j=1}^{r} A^{ij}(x) \frac{\partial H(x)}{\partial x^i} \frac{\partial H(x)}{\partial x^j} = 0, \quad LH(x) = 0. \tag{6.1}$$

This follows immediately from the Itô formula. Of course, a non-trivial first integral $H(x)$ exists only for degenerate diffusion processes. A diffusion process can have first integrals that are not smooth. For example, the process in $\mathbf{R}^1$, corresponding to the operator

$$L = \sin^2 x \frac{d^2}{dx^2} + \sin x \frac{d}{dx},$$

has a first integral $H(x) = k$ for $x \in [2\pi k, 2\pi(k+1))$.

A Markov process may have many first integrals. Any function of the first integrals will also be a first integral. If all the diffusion coefficients of the operator L vanish, then the process becomes a deterministic dynamical system, and (6.1) holds if

$$LH(x) = b(x) \cdot \nabla H(x) = 0, \quad x \in \mathbf{R}^r.$$

Consider now random perturbations of the process (X_t, P_x) corresponding to L. If $\sigma_0^2(x) = (A^{ij}(x))$ and $B(x) = (B^1(x), \ldots, B^r(x))$, the trajectories X_t can be described by the stochastic differential equation

$$dX_t = \sigma_0(X_t) \, dW_t + B(X_t) \, dt.$$

Let the perturbed process $(\tilde{X}_t^\epsilon, \tilde{P}_x^\epsilon)$ be defined by the perturbed equation

$$d\tilde{X}_t^\epsilon = \sigma_0(\tilde{X}_t^\epsilon) \, dW_t + B(\tilde{X}_t^\epsilon) \, dt + \sqrt{\epsilon}\sigma_1(\tilde{X}_t^\epsilon) \, d\tilde{W}_t + \epsilon b(\tilde{X}_t^\epsilon) \, dt;$$

here $\sigma_1(x)\sigma_1^*(x) = (a^{ij}(x))$, $b(x) = (b^1(x), \ldots, b^r(x))$, $\tilde{W}_t$ and W_t are independent Wiener processes in $\mathbf{R}^r$. Let us rescale the time: $X_t^\epsilon = \tilde{X}_{t/\epsilon}^\epsilon$. The generator of the new process $(X_t^\epsilon, P_x^\epsilon)$ is

$$L^\epsilon = \frac{1}{\epsilon} L + L_1, \quad L_1 = \frac{1}{2} \sum_{i,j=1}^{r} a^{ij}(x) \frac{\partial^2}{\partial x^i \partial x^j} + \sum_{i=1}^{r} b^i(x) \frac{\partial}{\partial x^i}.$$

Assume that the operator L_1 is uniformly elliptic with smooth bounded coefficients.

Let the non-perturbed process (X_t, P_x) have a smooth first integral $H(x)$. We assume that $H(x)$ is a generic function: it has a finite number of critical points, all of them are non-degenerate, and each level set of $H(x)$ has at most one critical point. Assume that $H(x) \geq 0$, $\lim_{|x| \to \infty} H(x) = \infty$ and $\min_x H(x) = 0$. Let $C(y) = \{x \in \mathbb{R}^r : H(x) = y\}$, $y \geq 0$. The set $C(y) \subset \mathbb{R}^r$ is compact and consists of a finite number $n(y)$ of connected components $C_i(y) : C(y) = \bigcup_{i=1}^{n(y)} C_i(y)$.

All the coefficients of the operator L vanish at the critical points of $H(x)$. Assume that if $x \in C_i(y)$ is not a critical point, then $\sum_{i,j=1}^{r} A^{ij}(x) e_i e_j \geq a(x) \cdot |e|^2$ for some $a(x) > 0$ and any $e = (e_1, \ldots, e_r)$, $e \cdot \nabla H(y) = 0$. This means that the process (X_t, P_x) is not degenerate at x if considered on the manifold $C_i(y)$. Then the process has a unique invariant density $m_y^i(x)$, $x \in C_i(y)$, on each connected component $C_i(y)$, if $C_i(y)$ contains no critical points of $H(x)$. If $C_i(y)$ contains a critical point z, the only invariant measure of (X_t, P_x) on $C_i(y)$ is concentrated at z.

Since the non-perturbed process has the first integral $H(x)$, one can introduce fast and slow components of the displacements of the perturbed process $(X_t^\epsilon, P_x^\epsilon)$: the fast motion along the level sets of the first integral and the slow motion in the transversal direction. The evolution of the slow component can be described, at least in a neighborhood of a non-singular level set $C_i(y)$, by the evolution of $H(X_t^\epsilon)$. The fast component near the manifold $C_i(y)$ has a distribution close to the invariant density $m_y^i(x)$ if $0 < \epsilon \ll 1$.

Consider, as we did in Sections 4 and 5, the evolution of the slow component. Applying the Itô formula, we have:

$$H(X_t^\epsilon) - H(x) = \frac{1}{\sqrt{\epsilon}} \int_0^t \nabla H(X_s^\epsilon) \cdot \sigma_0(X_s^\epsilon) \, dW_s + \frac{1}{\epsilon} \int_0^t LH(X_s^\epsilon) \, ds \quad (6.2)$$
$$+ \int_0^t \nabla H(X_s^\epsilon) \cdot \sigma_1(X_s^\epsilon) \, d\tilde{W}_s + \int_0^t L_1 H(X_s^\epsilon) \, ds.$$

Since $H(x)$ is a smooth first integral, condition (6.1) is fulfilled, and two integrals in the right-hand side of (6.3) having large factors vanish:

$$H(X_t^\epsilon) - H(x) = \int_0^t \nabla H(X_s^\epsilon) \cdot \sigma_1(X_s^\epsilon) \, d\tilde{W}_s + \int_0^t L_1 H(X_s^\epsilon) \, ds. \quad (6.3)$$

Consider now the graph Γ corresponding to the function $H(x)$ (see Section 5). Let Γ consist of the edges $I_1, \ldots, I_n$ and the vertices $O_1, \ldots, O_m$. Let $Y : \mathbb{R}^r \to \Gamma$

be the mapping defined in Section 5 for $r = 2$: $Y(x)$ is the point of Γ corresponding to the connected component of the level set $C(H(x))$ containing the point $x \in \mathbb{R}^r$. Consider the random process $Y_t^\epsilon = Y(X_t^\epsilon)$. Any interior point y of an edge I_i corresponds to a non-singular connected component $C_k(y)$ of a level set of $H(x)$. Then, using (6.3) and the ergodicity of the non-perturbed process on $C_k(y)$, one can prove that there exists a weak limit Y_t of the processes Y_t^ϵ. Inside the edge I_i, the process Y_t is a diffusion process governed by the operator

$$
\begin{aligned}
\bar{L}_i &= \frac{1}{2}\bar{a}_i(y)\frac{d^2}{dy^2} + \bar{b}_i(y)\frac{d}{dy}, \\
\bar{a}_i(y) &= \int_{C_k(y)} \sum_{l,j} a^{lj}(x)\frac{\partial H(x)}{\partial x^l}\frac{\partial H(x)}{\partial x^j}m_y^k(x)\,dx, \qquad (6.4) \\
\bar{b}(y) &= \int_{C_k(y)} L_1 H(x)m_y^k(x)\,dx.
\end{aligned}
$$

To give a full description of the limiting process, one should prove that it is Markovian and give the gluing conditions at the vertices. It is easy to check that the vertices corresponding to the extrema of $H(x)$ (exterior vertices) are inaccessible. Therefore, no gluing (boundary) conditions should be imposed at these vertices. The calculation of the gluing conditions at the interior vertices is a more complicated problem. I will just mention a trick which sometimes helps: Assume that the processes in $\mathbb{R}^r$ corresponding to the operators L and L_1 have the same invariant density $m(x)$. Then the processes $(X_\cdot^\epsilon, P_x^\epsilon)$ have, for any $\epsilon > 0$, the same invariant density $m(x)$, and one can calculate the invariant measure for the limiting process Y_t on the graph. As it was explained in the previous section, the gluing conditions at a vertex $O_k \subset \Gamma$ are described by a set of constants $\alpha_k, \beta_{k1}, \dots, \beta_{kl} > 0$, where $l = l(k)$ is the number of edges connected with O_k. These constants can be chosen in a unique way to have the prescribed invariant density. For example, one can use such an approach if

$$
L = \widehat{L} + B(x) \cdot \nabla, \qquad L_1 = \widehat{L}_1 + b(x) \cdot \nabla,
$$

where $\widehat{L}$ and $\widehat{L}_1$ are self-adjoint operators, and $B(x)$ and $b(x)$ are divergence free vector fields. The Lebesgue measure is invariant for L and L_1 in this case.

2. The non-perturbed process (or the dynamical system) may have several smooth first integrals $H_1(x), \dots, H_l(x); l < r$. Let $C(y) = \{x \in \mathbb{R}^r : H_1(x) = y_1, \dots, H_l(x) = y_l\}$, $y = (y_1, \dots, y_l) \in \mathbb{R}^l$. The set $C(y)$ may be empty for some $y \in \mathbb{R}^l$. Let $C(y)$ consist of $n(y) < \infty$ connected components: $C(y) = \bigcup_{k=1}^{n(y)} C_k(y)$, $y \in \mathbb{R}^l$. Assume that at least one of the first integrals, say $H_1(y)$, has the property $\lim_{|x| \to \infty} H_1(x) = \infty$. Then each $C_k(y)$ is compact.

We say that a point $x \in \mathbb{R}^r$ is non-singular, if the rank of the matrix $\left(\frac{\partial H_i(x)}{\partial x^j}\right)$, $1 \leq i \leq l$, $1 \leq j \leq r$, is equal to l. Assume that the non-perturbed process at a non-singular point $x \in \mathbb{R}^r$ is non-degenerate, if considered on $C(y)$, $y = (H_1(x), \ldots, H_l(x))$. This means that

$$\sum_{i,j} A^{ij}(x) e_i e_j \geq a(x) |e|^2, \quad a(x) > 0,$$

for any $e = (e_1, \ldots, e_r)$ such that $e \cdot \nabla H_k(x) = 0$, $k = 1, \ldots, l$. Then the process (X_t, P_x) has a unique invariant measure on each $C_k(y)$ consisting of non-singular points.

The collection of all connected components of the level sets $C(y)$, $y \in \mathbb{R}^l$, provided with the natural topology, is homeomorphic to a set Γ consisting of glued l-dimensional pieces. The interior points of those pieces correspond to the components $C_k(y)$ consisting of non-singular points. Let $Y : \mathbb{R}^r \to \Gamma$ be the mapping similar to the mapping Y defined in Section 5: $Y(x)$ is the point of Γ corresponding to the connected component of $C(H_1(x), \ldots, H_l(x))$, containing $x \in \mathbb{R}^r$. Then one can expect that the processes $Y_t^\epsilon = Y(X_t^\epsilon)$ converge weakly (in the space C_{0T} of continuous functions on $[0, T]$, $T < \infty$, with values in Γ) to a continuous process Y_t on Γ as $\epsilon \downarrow 0$. Inside an l-dimensional piece $\gamma_k \subset \Gamma$, the values of the first integrals can be used as the coordinates. In these coordinates the process Y_t is governed by the operator

$$\bar{L}_k = \frac{1}{2} \sum_{i,j=1}^{l} \bar{a}_k^{ij}(y) \frac{\partial^2}{\partial y^i \partial y^j} + \sum_{i=1}^{l} \bar{b}_k(y) \frac{\partial}{\partial y^i}.$$

The coefficients $\bar{a}_k^{ij}(y)$ and $\bar{b}_k^i(y)$ can be calculated by averaging $a^{ij}(x)$ and $b^i(x)$ with respect to the invariant densities of the non-perturbed process on the corresponding component of $C(Y)$ (compare with (6.4)).

To define the limiting process one should supplement the operators $\bar{L}_k$ governing the process inside the l-dimensonal pieces by some gluing conditions in the places where several l-dimensional pieces are glued together. These gluing conditions should be, in a sense, similar to the boundary conditions for the multi-dimensional processes.

3. When we considered perturbations of the dynamical system shown in Fig. 8, as well as the problems of this section, we assumed that the non-perturbed process (or the dynamical system) has a unique invariant measure on each connected component $C_k(y)$ of the level sets $C(y)$ of the first integral. In this case, the limit of the projection $Y(X_t^\epsilon)$ of the perturbed process X_t^ϵ on the graph, counting the connected components, turns out to be a Markov process. I want to explain here that, actually, the structure of the set of invariant measures for the non-perturbed process is important, but not the set of connected components. The non-degeneration assumption, or, in the case of system (5.1), the assumption that the vector field has no rest points besides the critical points of the first integral, lead

to uniqueness of a normalized invariant measure on each connected component of the level sets. Therefore, under these conditions, the graph describes the invariant measures for the non-perturbed system as well.

Consider an example: Let the non-perturbed process (X_t, P_x) be the process in the strip $\Pi = \{x = (x^1, x^2) \in R^2 : |x^1| \leq 1\}$ governed by the operator

$$L = \frac{1}{2} A(x) \frac{\partial^2}{(\partial x^1)^2} + B(x) \frac{\partial}{\partial x^1}$$

inside the strip and having normal reflection on the boundary $\partial \Pi = \{x \in R^2 : |x^1| = 1\}$. It is clear that the function $H(x) = x^2$ is a first integral for such a process. All the level sets $C(y) = \{(x^1, x^2) \in \Pi : x^2 = y, |x^1| \leq 1\}$ are unit intervals.

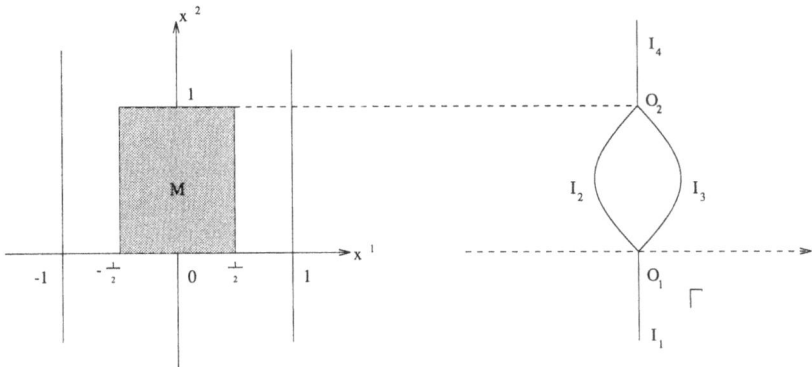

Figure 13

Assume that the coefficients $A(x)$, $B(x)$ are smooth and bounded and that $A(x) > 0$ everywhere in Π besides the set $K = \{(x^1, x^2) \in \Pi : |x^1| = \frac{1}{2}, 0 \leq x^2 \leq 1\}$. Let $\text{sign} B(x) = \text{sign} x^1$ for $x = (x^1, x^2) \in K$. Then the process (X_t, P_x) in Π has one normalized invariant measure μ_y on $C(y)$ for any $y \notin [0, 1]$, and two extreme normalized invariant measures μ_y^1 and μ_y^2 for $y \in [0, 1]$. The measure μ_y, $y \notin [0, 1]$, is concentrated on the set $\{(x^1, x^2) : -1 \leq x^1 \leq 1, x^2 = y\}$ and has a positive density with respect to $x^1 \in [-1, 1]$. This density can be calculated explicitly solving the corresponding stationary backward Kolmogorov equation. The measure μ_y^1 (μ_y^2) for $y \in [0, 1]$ has its support on $\{(x^1, x^2) : x^1 \in [-1, -\frac{1}{2}], x^2 = y\}$ ($\{(x^1, x^2) : x^1 \in [\frac{1}{2}, 1], x^2 = y\}$). The densities of these measures also can be calculated explicitly. Of course, any combination of the measures μ_y, μ_y^1, μ_y^2 with positive coefficients will be an invariant measure, as well. But such a combination will not be an extreme point of the cone of invariant measures.

Let $L_1 = \frac{1}{2} \sum_{i,j=1}^{2} a^{ij}(x) \frac{\partial^2}{\partial x^i \partial x^j} + \sum_{i=1}^{2} b^i(x) \frac{\partial}{\partial x^i}$.

The coefficients are assumed to be smooth and bounded, and $\sum_{i,j=1}^{2} a^{ij}(x)\lambda_i\lambda_j \geq a\sum_{1}^{2}\lambda_i$ for some $a > 0$. Assume for brevity that $(a^{ij}(x))$ is equal to the unit matrix for $x^2 = 0$ and $x^2 = 1$. Let the perturbed process $(\tilde{X}_t^\epsilon, \tilde{P}_x^\epsilon)$ be governed by the operator $L + \epsilon L_1$. Denote by (X_t^ϵ, P_x) the rescaled process: $X_t^\epsilon = \tilde{X}_{t/\epsilon}^\epsilon$; the operator $\frac{1}{\epsilon}L + L_1$ is the generator of (X_t^ϵ, P_x) inside Π.

Consider the evolution of the first integral $H(X_t^\epsilon) = X_t^{\epsilon,2}$ for $0 < \epsilon \ll 1$. Before $X_t^{\epsilon,2}$ moves a little, the first component $X_t^{\epsilon,1}$ approaches its limit distribution which is equal to an invariant measure on $C(X_t^{\epsilon,2})$. If $X_t^{\epsilon,2} \in [0,1]$, the limiting distribution depends on the position of the first component at the time of the last entrance before t to $[0,1]$ by the second component.

Using these arguments, under some mild additional assumptions on the behavior of the coefficients $A(x)$, $B(x)$ near the lines $x^2 = 0$ and $x^2 = 1$, one can prove the following result:

Consider the graph Γ (Fig. 13). One can use the coordinate $y = x^2$ as (local) coordinate on each edge. Define the mapping Y of a part of Π in Γ: $Y(x^1, x^2)$ is the point of I_1 or I_4 with the coordinate $y = x^2$ if $x^2 \notin [0,1]$; $Y(x^1, x^2)$ is the point of I_2 with the coordinate $y = x^2$ if $x^1 \in [-1, -\frac{1}{2}]$, $x^2 \in [0,1]$; and $Y(x^1, x^2)$ is the point of I_3 with the coordinate $y = x^2$ if $x^1 \in [\frac{1}{2}, 1]$, $x^2 \in [0,1]$. The mapping Y is not defined in "the unstable area" M shaded in Fig. 13. One can check that $P_x^\epsilon\{X_t^\epsilon \in M \text{ for some } t \in [0,T]\} \to 0$ as $\epsilon \downarrow 0$ for any $T < \infty$ and $x \in \Pi\backslash M$. If the trajectory X_t^ϵ starts at $x \in M$ then $P_x^\epsilon\{X_t^\epsilon \text{ visits M after time } \delta\} \to 0$ as $\epsilon \downarrow 0$ for any $\delta > 0$. Therefore, it is not important how $Y(x)$ is defined for $x \in M$. Let, for instance, $Y(x^1, x^2)$ be the point of I_2 with the coordinate $y = x^2$ if $(x^1, x^2) \in M$.

The processes $Y_t^\epsilon = Y(X_t^\epsilon)$, $X_0^\epsilon = x \notin M$, converge weakly as $\epsilon \downarrow 0$ in the space of continuous functions on $[0,T]$ with values in Γ to a diffusion process Y_t on Γ. The process Y_t is governed by an operator L_i inside each edge $I_i \subset \Gamma$, $i = 1, 2, 3, 4$, and by gluing conditions at O_1 and O_2. The operators L_i are defined as follows:

$$L_i = \frac{1}{2}a_i(y)\frac{d^2}{dy^2} + b_i(y)\frac{d}{dy};$$

$$a_i(y) = \int_{-1}^{1} a(x^1, y)\mu_y(dx^1) \text{ and } b_i(y) = \int_{-1}^{1} b(x^1, y)\mu_y(dx^1) \text{ for } i = 1, 4;$$

$$a_2(y) = \int_{-1}^{-1/2} a(x^1, y)\mu_y^1(dx^1), \quad b_2(y) = \int_{-1}^{-1/2} b(x^1, y)\mu_y^1(dx^1);$$

$$a_3(y) = \int_{1/2}^{1} a(x^1, y)\mu_y^2(dx^1), \quad b_3(y) = \int_{1/2}^{1} b(x^1, y)\mu_y^2(dx^1).$$

The gluing conditions at O_k, $k = 1, 2$, are the following: a function $u(z)$, $z \in \Gamma$, smooth inside the edges and continuous on Γ belongs to the domain of definition of the generator of the process Y_t on Γ if $\lim_{z \in I_i, z \to O_k} L_i u(z)$ is the same for all i, such that $I_i \sim O_k$ and

$$\sum_{i=2,3} \beta_{1i} \frac{du}{dy^i}(O_1) = \frac{du}{dy^1}(O_1), \quad \sum_{i=2,3} \beta_{2i} \frac{du}{dy^i}(O_2) = \frac{du}{dy^4}(O_2).$$

Here $\frac{d}{dy^i}$ means the differentiation in the coordinate $y = x^2$ along I_i. The constants β_{ki} are defined as follows:

$$\beta_{12} = \lim_{y \uparrow 0} \mu_y^1 \left(\left[-1, -\frac{1}{2} \right] \right), \quad \beta_{13} = \lim_{y \uparrow 0} \mu_y^2 \left(\left[\frac{1}{2}, 1 \right] \right),$$

$$\beta_{22} = \lim_{y \downarrow 1} \mu_y^1 \left(\left[-1, -\frac{1}{2} \right] \right), \quad \beta_{23} = \lim_{y \downarrow 1} \mu_y^2 \left(\left[\frac{1}{2}, 1 \right] \right).$$

The operators L_i and the gluing conditions define the limit process in a unique way.

Note, that if the trajectory X_t^ϵ starts at a point $x = (x^1, x^2) \in M$, then X_t^ϵ exits M very fast as $\epsilon \downarrow 0$, and $U^\epsilon(x) = P_x^\epsilon\{X_t^{\epsilon,1}$ hits $-\frac{1}{2}$ before $\frac{1}{2}\} \to u(x^1)$. The function $u(x^1)$ is the unique solution of the problem:

$$\frac{1}{2} A(x^1, x^2) \frac{d^2 u(x^1)}{(dx^1)^2} + B(x^1, x^2) \frac{du(x^1)}{(dx^1)} = 0, \quad -\frac{1}{2} < x^1 < \frac{1}{2},$$

$$u\left(-\frac{1}{2} \right) = 1, \quad u\left(\frac{1}{2} \right) = 0.$$

The problem considered here is similar to a problem studied in [FW2], where a process with a finite number states was considered as the fast component.

One should say, finally, that we could see the importance of the structure of stable extreme points of the cone of invariant measures, when perturbations of the system shown in Fig 11 were considered: There are two stable and two unstable equilibrium points for the dynamical system on each level set $C(y) = \{x \in \mathbb{R}^2 : H(x) = y\}$ for $H(B) < y < H(A)$. The graph in Fig 11b "counts", actually, the extreme measures concentrated at the stable equilibriums.

4. As it was explained in section 5, the gluing conditions at the vertex O have the form

$$\alpha L u(O) + \sum_{i \, I_i \sim O} \beta_i \frac{du}{dy^i}(O) = 0,$$

where α, β_i are nonnegative numbers, $\alpha + \sum_{i:I_i \sim O} \beta_i \neq 0$. The constants β_i, roughly speaking, describe the distribution between the edges, when the particle leaves the vertex. The constant α characterizes the delay at the vertex O. In the problems considered so far, the limiting process has no delay at the vertices, so that $\alpha = 0$. I will consider here an example of a dynamical system such that the limiting process has a delay at a vertex.

Let X_t be a dynamical system on two-dimensional torus T^2 preserving the area:

$$\dot{X}_t = b(X_t). \tag{6.5}$$

It was shown in [A], that "in the case of general position" the system has the following structure: there are a finite number N of loops (See Fig. 14) in T^2. Inside each loop the system behaves as a Hamiltonian system in a finite part of the plane. The trajectories outside of the loops are dense in the exterior of the loops. They form one ergodic class (See [A], [SKh]). There are two such loops in Fig. 14.

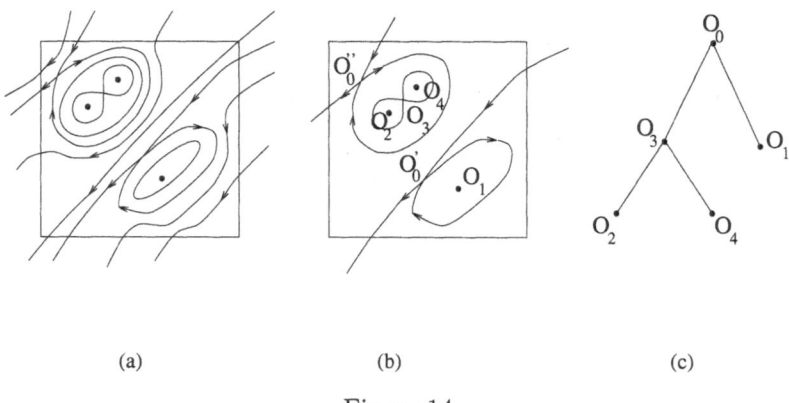

(a)   (b)   (c)

Figure 14

Consider now perturbations of system (6.5)

$$\dot{\tilde{X}}^\epsilon_t = b\left(\tilde{X}^\epsilon_t\right) + \sqrt{\epsilon}\dot{W}_t,$$

and let $X^\epsilon_t \doteq \tilde{X}^\epsilon_{t/\epsilon}$. Again, we can speak of a fast component of X^ϵ_t, which is the motion along the trajectories of the dynamical system, and of a slow component consisting of the motion in the direction transversal to the non-perturbed system. The exterior of the loops is one ergodic class for the fast motion so that all these points, if we want to describe the slow motion, should be glued together. Now, inside the loops, the system behaves as a Hamiltonian system in a part of the plane. One can use the values of the Hamiltonian as the slow coordinate. Without loss of generality, we can assume that there the Hamiltonian $H(x)$ has the same value, say zero, on all the loops. The graph Γ in Fig. 14c "counts", in a sense, the stable extreme invariant measures of our dynamical system. The point O_0

corresponds to the whole exterior of the loops. The edge O_0O_1 corresponds to the trajectories inside the loop containing the point O'_0. The edge O_0O_3 corresponds to the trajectories inside the loop containing O''_0 and outside the ∞-shape curve with the crossing point O_3. The edges O_3O_2 and O_3O_4 correspond to the trajectories inside the left and the right part of the ∞-shaped curve.

Let $Y : \mathrm{R}^2 \to \Gamma$ be the corresponding mapping. One can consider "the projection" Y_t^ϵ of the perturbed process on Γ: $Y_t^\epsilon = Y(X_t^\epsilon)$. The processes Y_t^ϵ should converge weakly to a Markov process Y_t on Γ. Inside the edges, the process Y_t is governed by differential operators, which can be calculated using the averaging procedure, exactly as in the case considered in section 5. The gluing conditions at O_3 have the same form as in section 5 as well. The points O_1, O_2, O_4 turn out to be inaccessible and no conditions should be imposed there. A new effect is observed at the vertex O_0. Since system (6.5) preserves the area, the vector field $b(x)$ is divergence free. Then the uniform distribution on T^2 is the (unique) invariant measure for the process X_t^ϵ on T^2 for any $\epsilon > 0$. One can calculate "the projection" of the uniform distribution on Γ. This measure on Γ will be the invariant measure for the limiting process Y_t. Now, we can choose the gluing condition at O_0 so that the process with this gluing condition has the prescribed invariant measure. It can be done in a unique way. Since the exterior of the loops has a positive area, the invariant measure of the point O_0 for the limiting process on Γ is positive. Thus the gluing condition at O_0 has the delay term: $\alpha > 0$. The gluing conditions with a delay appear also when perturbations of Markov processes with conservation laws are studied. We will consider in the next section a problem where the limiting process on a graph has a delay at some vertices as well.

5. Suppose the evolution of a physical system is described by a differential equation in R^2 perturbed by a small white noise:

$$\dot{\tilde{X}}_t^\epsilon = b\left(\tilde{X}_t^\epsilon\right) + \sqrt{\epsilon}\dot{W}_t, \ \ \dot{X}_0^\epsilon = x \in \mathrm{R}^2, \ 0 < \epsilon \ll 1. \tag{6.6}$$

Assume, that a domain $G \subset \mathrm{R}^r$ (critical domain) is singled out, so that the physical system is "alive" until $\tilde{X}_t^\epsilon \in G$, and the system "dies" when $\tilde{X}_t^\epsilon$ hits the boundary ∂G. Then, the expectation of the first exit time is a good characteristic of the vitality of the system. One can assume that the non-perturbed vector field $b(x)$ does not push the system to the boundary, and the death occurs because of the perturbations.

Suppose, we can control the system by adding an additional vector field $\beta(x)$ from a permissible set M to the right-hand side of (6.6). Since the exit occurs due to the noise, it is natural to assume, sometimes, that the control has the same order as the noise. This explains the statement of the following problem:

Consider a Hamiltonian system in R^2 (Fig. 15a). Let $H(x)$ be its Hamilton function. We assume that $H(x)$ is smooth enough, has a finite number of nondegenerate critical points and $\lim_{|x| \to \infty} H(x) = \infty$. Let the critical domain G be as in Fig. 15b, assume, for brevity that ∂G consists of 3 trajectories of the dynamical system: ∂G_1, ∂G_2, ∂G_3. The perturbed system (with the control), after time rescaling, has the form

$$\dot{X}_t^{\epsilon,\beta} = \frac{1}{\epsilon}\overline{\nabla}H\left(X_t^{\epsilon,\beta}\right) + \beta\left(X_t^{\epsilon,\beta}\right) + \dot{W}_t, \quad X_0^{\epsilon,\beta} = x \in \mathbb{R}^2. \tag{6.7}$$

Assume, that we can choose as a control $\beta(x)$ any vector field satisfying the conditions: $|\beta(x)| \leq K$ for some $K < \infty$; $\beta(x)$ is continuous everywhere besides at a finite number of the level sets $C(y) = \{x \in \mathbb{R}^r : H(x) = y\}$, where $\beta(x)$ has a simple discontinuity (it means, that the limits from both sides of the singular $C(y)$ exist and are continuous functions); moreover, assume that outside the curves $C(y)$, where $\beta(x)$ has jumps, the function $\beta(x)$ is smooth enough.

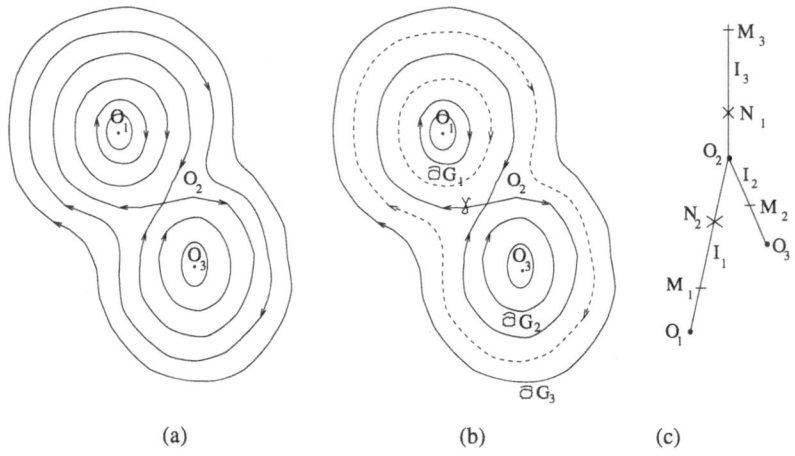

(a)   (b)   (c)

Figure 15

Let M be the set of all permissible controls. For each $\beta(\cdot) \in M$, we have the process $(X_t^{\epsilon,\beta}, P_x^{\epsilon,\beta})$ defined by (6.7). Denote by $\tau^{\epsilon,\beta}$ the first exit time

$$\tau^{\epsilon,\beta} = \min\{t : X_t^{\epsilon,\beta} \notin G\}.$$

Our goal is to find an admissible control, which makes the average lifetime of the system maximal. One can consider this problem for a fixed ϵ and write down a Bellman equation for $V^\epsilon(x) = \sup_{\beta \in M} E_x \tau^{\epsilon,\beta}$:

$$\frac{1}{2}\Delta V^\epsilon(x) + \frac{1}{\epsilon}\overline{\nabla}H(x) \cdot \nabla V^\epsilon(x) + K|\nabla V^\epsilon(x)| = -1, \quad x \in G, \tag{6.8}$$
$$V^\epsilon(x)|_{\partial G} = 0.$$

If $V^\epsilon(x)$ is the solution of (6.8), then the optimal control is $\hat{\beta}^\epsilon(x) = \nabla V^\epsilon(x) \cdot K|\nabla V^\epsilon(x)|^{-1}$ (see [FW1] and the references there). But problem (6.8) is rather

complicated. It was suggested in [FW1] to look for an asymptotic optimal control:
To calculate the main term of $E_x \tau^{\epsilon,\beta}$ as $\epsilon \downarrow 0$ for a given $\beta(x)$, and then to choose
$\beta(x)$ which makes the main term as large as possible. Such an asymptotically
optimal control will be not worse than any other control if ϵ is small enough. Such
a solution is acceptable in many problems, especially, if we take into account that
the value of ϵ is often unknown. This approach was used in [FW1, Chapter 8] for
an optimal stabilization problem in the large deviation case.

For a given $\beta(x)$, using the arguments similar to those used in section 5,
one can prove that the slow component of the process $(X_t^{\epsilon,\beta}, P_x^{\epsilon,\beta})$ converges to
a diffusion process Y_t^β on graph Γ, corresponding to the Hamiltonian $H(x)$. The
points M_1, M_2, M_3 in Fig. 15c correspond to different parts ∂G_1, ∂G_2, ∂G_3 of
the boundary of the critical domain G. On each edge I_i, $i = 2,3$, the operators
L_i are defined by averaging. For example, L_1 is defined by averaging over the
components $C_1(y)$ of the level set $C(y)$, contained in the left part of the ∞-shaped
curve γ (see Fig. 15b):

$$L_1 = \frac{a_1(y)}{2}\frac{d^2}{dy^2} + (b_1(y) + \bar{\beta}_1(y))\frac{d}{dy}, \quad T_1(y) = \int_{C_1(y)} \frac{dl}{|\nabla H(x)|},$$

$$a_1(y) = \frac{1}{T_1(y)}\int_{C_1(y)} |\nabla H(x)|dl, \qquad b_1(y) = \frac{1}{2T_1(y)}\int_{C_1(y)} \frac{\Delta H(x)}{|\nabla H(x)|}dl,$$

$$\bar{\beta}_1(y) = \frac{1}{T_1(y)}\int_{C_1(y)} \frac{\nabla H(x) \cdot \beta(x)}{|\nabla H(x)|}dl.$$

The operators L_i should be supplemented by the gluing conditions at the vertices
corresponding to saddle points of $H(x)$. O_2 in Fig. 15 is such a vertex: a continuous
function $u(y)$ on Γ belongs to the domain of definition of the generator for the
limiting process on Γ, if it is smooth inside the edges and if

$$\sum_{i=1}^{3} \beta_{2i}\frac{du}{dy^i}(O_2) = 0,$$

where $\frac{d}{dy^i}$ means the differentation along I_i with respect to the natural coordinate
y on I_i, growing with the distance from O_2; $\beta_{2i} = \int_{\gamma_i} |\nabla H(x)|dx$, where γ_1 is the
left part of γ, γ_2 is the right part of γ and $\gamma_3 = \gamma$.

The convergence of the slow component of $(X_t^{\epsilon,\beta}, P_x^{\epsilon,\beta})$ implies that $E_x \tau^{\epsilon,\beta} \to$
$E_y \tau^\beta$, where $\tau^\beta = \min\{t : Y_t \notin \hat{\Gamma}\}$, $\hat{\Gamma}$ is the part of Γ containing O_2 and bounded
by the points M_1, M_2, M_3. Thus the asymptotically optimal control should max-
imize $E_y \tau^\beta$, $y \in \hat{\Gamma}$.

The last problem is one-dimensional. To solve it, we should consider the counterpart of the problem (6.8) in $\hat{\Gamma}$:

$$\frac{a_i(y^i)}{2}\frac{d^2\hat{V}(y^i)}{(dy^i)^2} + b_i(y^i)\frac{d\hat{V}(y^i)}{dy^i} + K\cdot d_i(y^i)\cdot\left|\frac{d\hat{V}(y^i)}{dy^i}\right| = -1,$$

$$y^i \in I_i \cap \hat{\Gamma},\ \ i = 1,2,3;$$

$$\hat{V}(M_1) = \hat{V}(M_2) = \hat{V}(M_3) = 0;$$

$$\sum_{i=1}^{3} \beta_{2i}\frac{d\hat{V}(O_2)}{dy^i} = 0.$$

(6.9)

Here β_{2i} is as before, and

$$a_i(y) \ = \ \frac{1}{T_i(y)}\int_{C_i(y)}|\nabla H(x)|dl, \quad b_i(y) \ = \ \frac{1}{2T_i(y)}\int_{C_i(y)}\frac{\Delta H(x)}{|\nabla H(x)|}dl,$$

$$T_i(y) \ = \ \int_{C_i(y)}\frac{dl}{|\nabla H(x)|} \qquad d_i(y) \ = \ \frac{1}{T_i(y)},$$

where $C_i(y)$ are corresponding components of the level set $C(y)$. Problem (6.9) has a unique solution continuous in $\hat{\Gamma}$, and this solution can be written down explicitly. In general, it can be reduced to the solution of a system of linear algebraic equations. If $\hat{V}(y)$, $y \in \hat{\Gamma}$, is the solution of (6.9), the asymptotically optimal control $\hat{\beta}(x)$, $x \in G$, has the form

$$\hat{\beta}(x) = K\frac{\nabla H(x)}{|\nabla H(x)|}\cdot \mathrm{sign}\hat{V}'\left(Y(x)\right),$$

where Y is the mapping from $\mathbb{R}^2$ to Γ defined before. So, the asymptotically optimal control consists of pushing the trajectory to one of finite number of "special" trajectories with maximal possible speed. The "special" trajectories are defined by the zeros of $V'(y)$, $y \in \Gamma'$. For example, for the system shown on Fig. 15, one or two such trajectories can exist. They are shown by the dotted lines; the points $N_1, N_2 \in \Gamma$ correspond to these trajectories. This problem is considered in [DF].

7 Diffusion Processes and PDE's in Narrow Branching Tubes

Consider a connected graph Γ in $\mathbb{R}^r$ with vertices $O_1, \ldots, O_m$ and edges $I_1, \ldots, I_n$. Let G^ϵ be a domain consisting of narrow tubes surrounding the edges $I_i \subset \Gamma$ and of small neighborhoods $\mathcal{E}_k$ of the vertices $O_k \subset \Gamma$ (see Fig. 16).

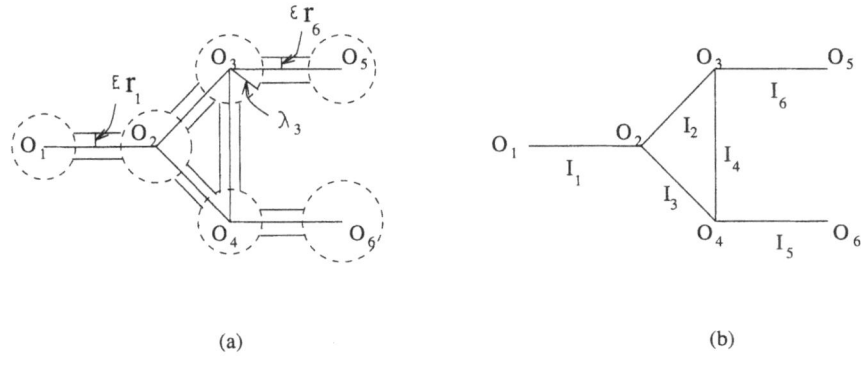

(a)   (b)

Figure 16

Assume for brevity that the cross-sections of the tubes are $(r-1)$-dimensional balls with centers on Γ, and that the neighborhoods of the vertices are balls in $\mathbb{R}^r$. Let ϵr_i be the radius of the tube surrounding I_i and $\varrho_j \lambda_j = \varrho_j \lambda_j(\epsilon)$ the radius of the ball $\mathcal{E}_j$ centered at O_j. Here ϱ_j, r_j are positive constants, and $\lim_{\epsilon \downarrow 0} \lambda_j(\epsilon) = 0$, $j = 1, \ldots, m$.

Consider the problem:

$$
\begin{aligned}
\frac{\partial u^\epsilon(t,x)}{\partial t} &= \frac{D}{2} \Delta u^\epsilon(t,x), \quad t > 0, \ x \in G^\epsilon \\
u^\epsilon(0,x) &= g(x), \quad \left. \frac{\partial u^\epsilon(t,x)}{\partial n(x)} \right|_{t>0, x \in \partial G^\epsilon} = 0,
\end{aligned}
\tag{7.1}
$$

where $g(x)$, $x \in \mathbb{R}^r$ is a continuous bounded function, and $n(x)$ is the normal to the boundary at $x \in \partial G^\epsilon$. We are interested in the behavior of the solution of (7.1) as $\epsilon \downarrow 0$.

Denote by $Y(x)$, $x \in G^\epsilon$, the point of Γ closest to x. If such a point is not unique take as $Y(x)$ any of them. One can expect that a function $\bar{u}(t,y)$, $y \in \Gamma$, exists such that

$$
\sup_{x \in G^\epsilon} |u^\epsilon(t,x) - \bar{u}(t, Y(x))| \to 0 \text{ as } \epsilon \downarrow 0.
$$

Moreover, since the equation (7.1) and the boundary conditions are invariant with respect to the shifts along each of the tubes and with respect to the reflection in any plane perpendicular to the edge, we can expect that the function $\bar{u}(t, y)$, $y \in \Gamma$, is a solution of the following problem on the graph:

$$\frac{\partial \bar{u}(t, y)}{\partial t} = \frac{D}{2} \frac{\partial^2 \bar{u}(t, y)}{\partial y^2}, \quad t > 0, \ y \in \Gamma \setminus \{O_1, \ldots, O_m\}, \tag{7.2}$$
$$\bar{u}(0, y) = g(y).$$

But the problem (7.2) has many solutions. To single out one solution we should add to (7.2) some gluing (boundary) conditions at the vertices.

From the probabilistic point of view this means, that if $(X_t^\epsilon, P_t^\epsilon)$ is the process in G^ϵ corroponding to the problem (7.1), then the projection $Y_t^\epsilon = Y(X_t^\epsilon)$ of X_t^ϵ on the graph converges as $\epsilon \downarrow 0$ to a diffusion process Y_t on Γ. Because of the invariance properties of problem (7.1) mentioned above, it is natural to expect that the process Y_t is governed inside the edges by the operator $\frac{D}{2} \frac{d^2}{dy^2}$. To define the limiting process for all $t \geq 0$, one should describe the behavior of the process Y_t at the vertices. This description is equivalent to the gluing conditions for (7.2).

In the case $\lambda_k(\epsilon) = \epsilon$, problem (7.1) was considered in [FW2]. It was shown there that the processes Y_t^ϵ converge to a continuous Markov process on the graph. The limiting process is governed by the operator $\frac{D}{2} \frac{d^2}{dy^2}$ inside the edges, and by gluing conditions

$$\sum_{i:I_i \sim O_k} r_i^{r-1} \frac{d\bar{u}}{dy^i}(O_k) = 0, \quad k \in \{1, \ldots, n\} \tag{7.3}$$

at the vertices. Here $I_i \sim O_k$ means that O_k is an end of the edge I_i; $\frac{d}{dy^i}$ denotes the differentiation in the natural coordinate on $I_i \sim O_k$, which is the distance of a point from O_k.

Condition (7.3) means that the limiting process has no delay at the vertices. If just one edge is connected with a vertex O_k (points O_1, O_5, O_6 in Fig. 16b) then (7.3) means that the process Y_t has at O_k instantaneous reflection. Problem (7.2–7.3) has a unique solution, which describes the limit of $u^\epsilon(t, x)$ as $\epsilon \downarrow 0$.

The proof of this statement, roughly speaking, consists of two parts: First, we prove that the limiting process is a continuous Markov process on Γ. This is, actually, a corollary of the fact, that the trajectories X_t^ϵ, starting at any $x \in \mathcal{E}_j$, mix in $\mathcal{E}_j$ before they leave the δ-neighborhood of O_j with probability close to one as $\epsilon \downarrow 0$. Here δ is a small, but independent of ϵ, positive number. And, second, since all possible gluing conditions for continuous Markov processes are known (see Section 5 and [FW2]), we should choose among them the conditions describing the limit process. This choice can be done if we note that the uniform distribution in G^ϵ is the unique invariant measure of the process X_t^ϵ for any $\epsilon > 0$. Using this fact one can easily calculate the normalized invariant measure for the limiting process, and then choose the gluing conditions so that the limiting process has the prescribed invariant density.

If for some $k \in \{1, \ldots, m\}$, $\lambda_k(\epsilon)\epsilon^{-2} \to \infty$ as $\epsilon \downarrow 0$, but $\lambda_k(\epsilon)\epsilon^{-\frac{r-1}{r}} \to 0$, then the invariant measure of the vertex O_k for the limiting process is still equal to zero. Thus the limiting process spends at O_k time zero, and the gluing (boundary) condition at such a point is again given by (7.3).

But the situation is different if $\lambda_k(\epsilon) \sim \epsilon^{\frac{r-1}{r}}$ as $\epsilon \downarrow 0$ or if $\lambda_k(\epsilon) \gg \epsilon^{\frac{r-1}{r}}$. Consider, first, the case $\lim_{\epsilon \downarrow 0} \lambda_k(\epsilon)\epsilon^{-\frac{r-1}{r}} = 1$. Of course, since the gluing conditions at O_k are of a local nature, it is sufficient to consider not the whole domain G^ϵ but just the part G^ϵ_δ of G^ϵ in a δ-neighborhood of O_k, where δ is a small positive number (Fig. 17a).

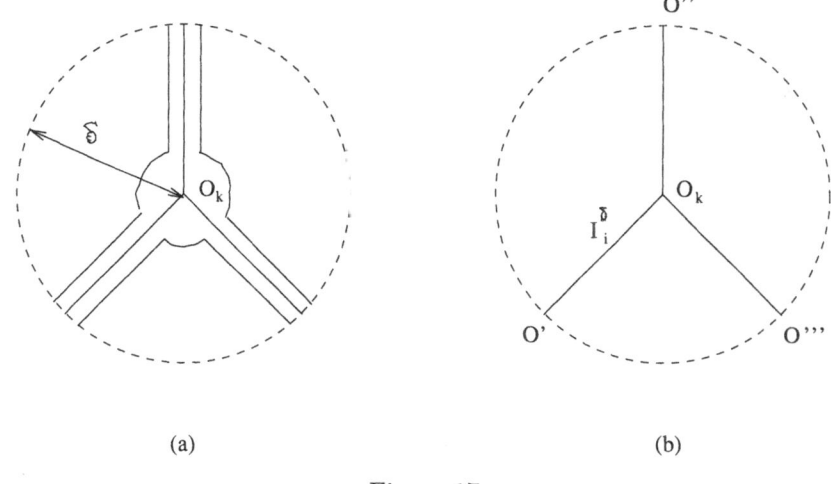

Figure 17

Let the process have the instantaneous normal reflection on the boundary of G^ϵ_δ. It is easy to see, that the normalized invariant measure μ of the limiting process in the phase space $\Gamma_\delta = \Gamma \cap \{x : |x - O_k| \leq \delta\}$ (see Fig. 17b), under our assumption, has a positive mass at O_k and has a constant density on each of edges I^δ_i connected with O_k:

$$\mu(O_k) = \frac{\varrho^r_k V_r}{V_r \varrho^r_k + \delta \sum_{i:I_i \sim O_k} r_i^{r-1} V_{r-1}},$$

$$m_i = \frac{d\mu}{dx}\Big|_{I^\delta_i} = \frac{r_i^{r-1} V_{r-1}}{V_r \varrho^r_k + \mathcal{E} \sum_{i:I_i \sim O_k} r_i^{r-1} V_{r-1}},$$

(7.4)

where V_r is the volume of a unit ball in $\mathbf{R}^r$.

$$\frac{V_{r-1}}{V_r} = \frac{\Gamma\left(\frac{r+2}{2}\right)}{\Gamma\left(\frac{1}{2}\right) \Gamma\left(\frac{r+1}{2}\right)}.$$

Formulas (7.4) follow from the fact that uniform distribution is invariant for the pre-limiting process in G_δ^ϵ, and that the measure μ is the projection of the normalized uniform distribution in G_δ^ϵ on Γ_δ.

Consider now the process (Y_t, P_y) on Γ_δ, governed by the operator $\frac{D}{2}\frac{d^2}{dy^2}$ inside each edge and having the instantaneous reflection at O', O'', O''' (see Fig. 17b), and satisfying at O_k the gluing condition

$$\sum_{i:I_i\sim O_k} \beta_{ki}\frac{du}{dy^i}(O_k) + \frac{\alpha_k D}{2}u''(O_k) = 0.$$

To calculate the invariant distribution, we need the operator A^* adjoint to the generator A of the process (Y_t, P_y); A^* is defined on measures. It is sufficient to calculate A^* for measures μ having a smooth density $m_i(y)$ inside each edge $I_i^\delta \subset \Gamma_\delta$ and a positive mass $\mu(\{O_k\})$ at O_k. Using the integration by parts and taking into account the boundary and gluing conditions for the generator A, we have

$$\int_{\Gamma_\delta} \frac{D}{2}u''(y)\mu(dy) = \frac{D}{2}u''(O_k)\mu(\{O_k\}) + \sum_{i:I_i\sim O_k}\int_{I_i^\delta}\frac{D}{2}u''(y)m_i(y)dy =$$

$$= -\frac{1}{\alpha_k}\sum_{i:I_i\sim O_k}\beta_{ki}\frac{du}{dy^i}(O_k)\mu(\{O_k\}) + \sum_{i:I_i\sim O_k}\int_{I_i^\delta}\frac{D}{2}m_i''(y)u(y)dy+ \qquad (7.5)$$

$$+\frac{D}{2}\sum_{i:I_i\sim O_k}(m_i(y)u'(y) - u(y)m_i'(y))\bigg|_{O_k}^{O^{(i)}},$$

where $O^{(i)}$ is the end of I_i^δ different from O_k. This equality implies that the densities $m_i(y)$ inside each I_i^δ satisfy equation $m_i''(y) = 0$, $y \in I_i$, and the boundary condition $m_i'(O^{(i)}) = 0$. Thus $m_i(y) = m_i = \text{const}$ for $y \in I_i^\delta$. It follows also from (7.5), that the invariant measure should satisfy the equality

$$\sum_{i:I_i\sim O_k} m_i u_i'(O_k) = \frac{\mu(\{O_k\})}{\alpha_k}\sum_{i:I_i\sim O_k}\beta_{ki}u_i'(O_k)$$

for any $u_i'(O_k)$, $i : I_i \sim O_k$. The last condition implies that $\mu(\{O_k\}) = C\alpha_k$, $m_i = C\beta_{ki}$, where C is a positive constant. Defining the constant from normalization condition $\mu(\Gamma_\delta) = 1$, we have:

$$\mu(\{O_k\}) = \frac{\alpha_k}{\alpha_k + \delta\sum\limits_{i:I_i\sim O_k}\beta_{ki}}, \quad m_i = \frac{\beta_{ki}}{\alpha_k + \delta\sum\limits_{i:I_i\sim O_k}\beta_{ki}}. \qquad (7.6)$$

Comparing (7.4) and (7.6), we have the gluing condition at O_k for the limiting process Y_t on Γ:

$$\sum_{i:I_i \sim O_k} r_i^{r-1} \frac{du}{dy^i}(O_k) + \frac{\varrho_k^r \Gamma(\frac{1}{2})\Gamma(\frac{r+1}{2})}{\Gamma(\frac{r+2}{2})} \frac{d^2u}{dy^2}(O_k) = 0. \tag{7.7}$$

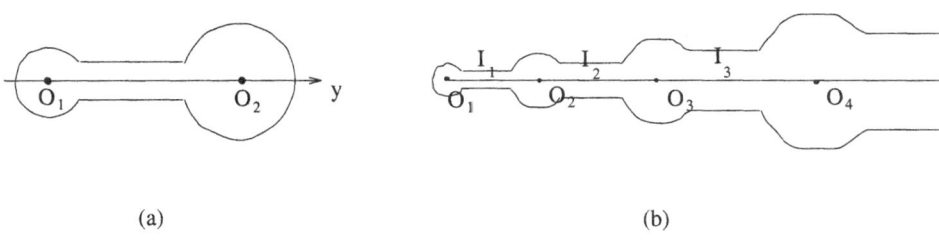

(a)   (b)

Figure 18

Recall that $\frac{d^2u}{dy^2}(O_k)$ is independent of i: just such functions $u(y)$, $y \in \Gamma$, belong to the domain of the generator A. In particular, if Γ consists of just one edge I (see Fig. 18a), $D = 1$, $r = 3$, and $\lim_{\epsilon \downarrow 0} \epsilon^{2/3} \lambda_j(\epsilon) = 1$ for $j = 1, 2$, the limiting process on the edge $O_1 O_2$ will be the Wiener process with the boundary conditions

$$-r_1^2 \frac{du}{dy}(O_1) + \frac{4}{3}\varrho_1^3 u''(O_1) = 0, \qquad r_1^2 \frac{du}{dy}(O_2) + \frac{4}{3}\varrho_2^3 u''(O_2) = 0.$$

The limiting process has delays at O_1 and at O_2.

Now let $\lambda_k(\epsilon) \gg \epsilon^{\frac{r-1}{r}}$ as $\epsilon \downarrow 0$. Then the point O_k becomes a trap for the limiting process Y_t: if the trajectory Y_t enters the vertex O_k, it stays there forever. The gluing condition at O_k then has the form:

$$\frac{d^2u}{dy^2}(O_k) = 0. \tag{7.8}$$

Formally, one can derive this from (7.7): if $\varrho_k = \infty$, then condition (7.7) should be replaced by (7.8). It is not difficult to give a rigorous proof of (7.8).

One should say, that the convergence of $Y(X_t^\epsilon)$ to the process Y_t on the graph holds only on finite time intervalls. Sooner or later, the process X_t^ϵ leaves the ball $\mathcal{E}_k$ for any $\epsilon > 0$, even if $\lambda_k(\epsilon) \gg \epsilon^{\frac{r-1}{r}}$, while Y_t never leaves the vertex when (7.8) holds. If $\lambda_k(\epsilon) \gg \epsilon^{\frac{r-1}{r}}$, the exit of X_t^ϵ from $\mathcal{E}_k$ during a finite time intervall is related with the large deviations. One can observe here the metastability effect (sublimiting distribution – in the terminology of [FW1]): The limiting distributions of $X_{t(\epsilon)}^\epsilon$ as $\epsilon \downarrow 0$ for different functions $t(\epsilon)$, $t(\epsilon) \to \infty$ as $\epsilon \downarrow 0$, can be different. This implies certain asymptotic behavior of the solutions of the corresponding parabolic equations.

Remark. The proof of the convergence of $Y_t^\epsilon = Y(X_t^\epsilon)$ to the process Y_t on the graph with the gluing (boundary) conditions (7.3) for the case $\lambda(\epsilon) \sim \epsilon$ was given in [FW2], using the martingale problem approach. This approach mixes together the proof of the Markov property for Y_t and the calculation of the gluing conditions. The same proof can be used for the case $\lambda_k(\epsilon) \ll \epsilon^{(r-1)/r}$. If $\lambda_k(\epsilon) \sim \epsilon^{(r-1)/r}$, one should make some additions to that proof: first, the exact asymptotics of the exit time from a small neighborhood of the vertex O_k should be calculated, and second, Theorem 4.1 from [FW2] should be extended to the case of vertices with delay. This was done in [BF].

We assumed that the radii of all tubes in G^ϵ have the same order as $\epsilon \downarrow 0$. If the radii have different order as $\epsilon \downarrow 0$, one can observe such an effect: Let, for example, G^ϵ be as shown in Fig. 18b, let $r_i(\epsilon)$ be the radius of the tube between O_i and O_{i+1} and $\lambda_k(\epsilon)$ the radius of the sphere $\mathcal{E}_k$ centered at O_k. Assume, that for any $i = 1, 2, \ldots$,

$$r_i(\epsilon) \ll [\lambda_{i+1}(\epsilon)]^{\frac{r}{r-1}} \sim r_{i+1}(\epsilon), \quad \epsilon \downarrow 0.$$

Then, asymptotically as $\epsilon \downarrow 0$ each ball is passable only form the left to the right. The long time effect is similar to the existence of a drift directed to the right. Consider now problem (7.1) in a narrow strip in R^2 with a variable cross-section.

Let $h(x)$, $k(x)$, $x \in \mathrm{R}^1$, be smooth positive functions, and

$$G^\epsilon = \{(x, y) \in \mathrm{R}^2 : -\epsilon k(x) \le y \le \epsilon h(x)\}.$$

Let $(X_t^\epsilon, Y_t^\epsilon; P_{x,y}^\epsilon)$ be the Wiener process in G^ϵ with normal reflection on the boundary. Again, one can expect that X_t^ϵ, $0 \le t \le T$, converges as $\epsilon \downarrow 0$ to an one-dimensional diffusion process. But now, due to the curvature of the boundary, an additional drift appears in the limiting process. Let $V(x) = \log[h(x) + k(x)]$, and consider the process in R^1 corresponding to the operator

$$I = \frac{1}{2}\frac{d^2}{dx^2} + V'(x)\frac{d}{dx}.$$

It turns out that X_t^ϵ, $0 \le t \le T$, converges to the diffusion process governed by I as $\epsilon \downarrow 0$. To prove this convergence, let us write down the stochastic differential equation for the process $(X_t^\epsilon, Y_t^\epsilon)$:

$$\begin{aligned}
dX_t^\epsilon &= dW_t^1 + \gamma_1^\epsilon(X_t^\epsilon, Y_t^\epsilon)\, dL_t^\epsilon, \\
dY_t^\epsilon &= dW_t^2 + \gamma_2^\epsilon(X_t^\epsilon, Y_t^\epsilon)\, dL_t^\epsilon,
\end{aligned} \tag{7.9}$$

where W_t^1, W_t^2 are independent Wiener processes in R^1, and L_t^ϵ is the local time for the process $(X_t^\epsilon, Y_t^\epsilon)$ on ∂G^ϵ; $\gamma^\epsilon(x, y) = (\gamma_1^\epsilon(x, y), \gamma_2^\epsilon(x, y))$ is the unit inward normal to ∂G^ϵ:

$$\gamma^\epsilon(x,y) \;=\; \frac{1}{\sqrt{1+\epsilon^2(h'(x))^2}}\,(\epsilon h'(x),-1) \quad \text{for } y = \epsilon h(x),$$

$$\gamma^\epsilon(x,y) \;=\; \frac{1}{\sqrt{1+\epsilon^2(k'(x))^2}}\,(\epsilon k'(x),1) \quad \text{for } y = -\epsilon k(x).$$

We see that $\gamma_2^\epsilon(x,y)$ is close to $-\mathrm{sign}(y)$, $|\gamma_1^\epsilon(x,y) - \epsilon h'(x)| = o(\epsilon)$ for $y = \epsilon h(x)$, and $|\gamma_1^\epsilon(x,y) - \epsilon k'(x)| = o(\epsilon)$ for $y = -\epsilon k(x)$ as $\epsilon \downarrow 0$.

Since Y_t^ϵ hits the boundary ∂G^ϵ more and more often as $\epsilon \downarrow 0$, the local time L_t^ϵ has the order ϵ^{-1} as $\epsilon \downarrow 0$. This means that system (7.9) has fast (Y_t^ϵ and L_t^ϵ) and slow (X_t^ϵ) components. As we have already seen earlier, the fast component with frozen slow component should be considered first in order to study such a system. Then the slow component can be approximated by the process with characteristics averaged in the fast variables.

The fast motion is described by the second of equations (7.9). If the slow variable in this equation is frozen, we have

$$d\bar Y_t^{\epsilon,x} = dW_t^2 - \mathrm{sign}(\bar Y_t^{\epsilon,x}) \cdot \chi_{-\epsilon a,\epsilon b}(\bar Y_t^{\epsilon,x})\,d\bar L_t^{\epsilon,x}, \tag{7.10}$$

where $\chi_{-\epsilon a,\epsilon b}$ is the indicator function of the set consisting of two points $-\epsilon a$ and ϵb, $a = k(x)$, $b = h(x)$. Equation (7.10) describes the Wiener process in $[\epsilon a, \epsilon b]$ with instantaneous reflection at the ends of the interval; $x \in \mathbf{R}^1$ is a fixed parameter.

To calculate the behavior of the local time $\bar L_t^{\epsilon,x}$ consider an auxiliary boundary problem:

$$u''(y) = \frac{\alpha + \beta}{a+b}, \quad -a < y < b, \quad u'(-a) = -\alpha, \quad u'(b) = \beta. \tag{7.11}$$

The right hand side $A = \frac{\alpha+\beta}{a+b}$ of equation (7.11) is chosen so that problem (7.11) is solvable. The solution is unique up to an additive constant.

Applying the Itô formula to $v^\epsilon(y) = \epsilon u(\epsilon^{-1}y)$, we obtain:

$$v^\epsilon(\bar Y_t^{\epsilon,x}) - v^\epsilon(\bar Y_t^{\epsilon,x}) = \int_0^t u'(\epsilon^{-1}\bar Y_s^{\epsilon,x})\,dW_s^2 + \frac{At}{2\epsilon} + \int_0^t l(\bar Y_t^{\epsilon,x})\,d\bar L_s^{\epsilon,x}, \tag{7.12}$$

where $l(y) = -\alpha$ for $y = -\epsilon a$ and $l(y) = \beta$ for $y = \epsilon b$. Equality (7.12) implies:

$$\sup_{0 \le t \le T} \left| \epsilon \int_0^t l(\bar Y_s^{\epsilon,x})\,d\bar L_s^{\epsilon,x} - \frac{\alpha+\beta}{2(a+b)}t \right| \to 0 \text{ as } \epsilon \downarrow 0, \tag{7.13}$$

say, in probability. Choosing $\alpha = k'(x)$, $\beta = h'(x)$ and recalling the behavior of $\gamma_1^\epsilon(x,y)$ for $\epsilon \downarrow 0$, one can derive from (7.13), that the term with the local time in the first of equations (7.9) is equivalent to the drift

$$\frac{1}{2} \cdot \frac{h'+k'}{h+k}(X_t^\epsilon)\,dt$$

as $\epsilon \downarrow 0$. Thus the component X_t^ϵ converges to the one-dimensional diffusion process $\bar{X}_t$ as $\epsilon \downarrow 0$:

$$d\bar{X}_t = dW_t^1 + b(\bar{X}_t)\,dt, \quad b(x) = \frac{1}{2}\frac{d}{dx}\ln(h(x) + k(x)).$$

It is not difficult to make these arguments rigorous (see [BF]).

We now mention a generalization of this problem for multidimensional processes. Let D be a set in $\mathbf{R}^d \times \mathbf{R}^m$ such that

$$\mathbf{R}^d \times \{0\} \subset D,$$

and let for any $x \in \mathbf{R}^d$ the set

$$D_x = \{y \in \mathbf{R}^m : (x, y) \in D\}$$

be a bounded connected domain. Assume that D has a smooth boundary and that the normal $n(x, y)$ to ∂D at any point $(x, y) \in \partial D$ is not parallel to $\mathbf{R}^d$. The domain D can be a tube in $\mathbf{R}^{d+m}$, if $d = 1$, or a layer, if $d > 1$ (see Fig. 19 a,b).

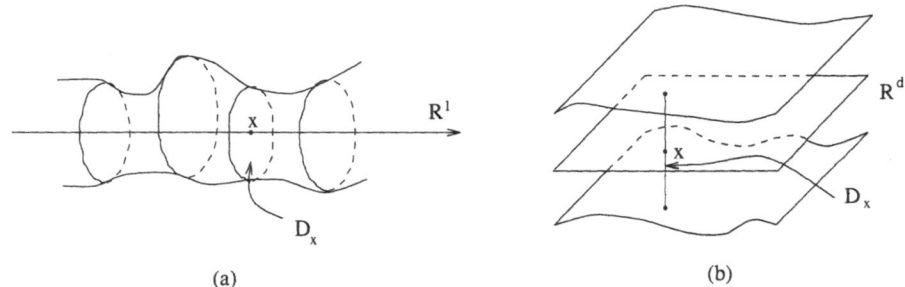

(a)   (b)

Figure 19

Let $D^\epsilon = \{(x, y) \in \mathbf{R}^{d+m} : (x, y\epsilon^{-1}) \in D\}$. Consider the Wiener process $(X_t^\epsilon, Y_t^\epsilon)$ in D^ϵ with normal reflection on ∂D^ϵ. Denote by $V(x)$ the volume of D_x in $\mathbf{R}^m$. Arguments similar to those used in the case of a strip in $\mathbf{R}^2$ allow to prove the following result:

Theorem 7.1 *The processes X_t^ϵ, $0 \le t \le T$, converge weakly in the space of continuous functions on $\mathbf{R}^d$ to the diffusion process governed by the operator*

$$\frac{1}{2}\Delta + \frac{1}{2}\nabla(\log V(x)) \cdot \nabla.$$

The process $(X_t^\epsilon, Y_t^\epsilon)$ can be described by equations (7.9), but now W_t^1, W_t^2 are Wiener processes in $\mathbf{R}^d$ and $\mathbf{R}^m$, respectively; $\gamma_1^\epsilon(x, y)$ is the projection of the unit inward normal to ∂D^ϵ on $\mathbf{R}^d$, and $\gamma_2^\epsilon(x,y)$ is the projection on $\mathbf{R}^m$. It is easy to see that $\lim_{\epsilon\downarrow 0}|\gamma_2^\epsilon(x,y)| = 1$, $\lim_{\epsilon\downarrow 0}\epsilon^{-1}\gamma_1^\epsilon(x,y) = \gamma_1^1(x,y)$.

To explain where the drift $\nabla\log V(x)$ comes from consider the boundary problem

$$
\begin{aligned}
\Delta u(y) &= A, \quad y \in D_x \subset \mathbf{R}^m, \\
\frac{\partial u(y)}{\partial \bar{n}(y)}\bigg|_{y\in\partial D_x} &= C(y),
\end{aligned}
\tag{7.14}
$$

where $\bar{n}(y)$ is the unit inward normal to ∂D_x, A is a constant, $C(y)$ is a smooth function on ∂D_x, and $x \in \mathbf{R}^d$ is a fixed parameter. Problem (7.14) is solvable if and only if

$$
A = \frac{1}{V(x)}\int_{\partial D_x} C(y)\,ds,
$$

where ds is the surface element on ∂D_x. Applying the Itô formula to $\epsilon u(\epsilon^{-1}Y_t^\epsilon)$, one can calculate the limiting behavior of the term with local time in the stochastic equation for X_t^ϵ. Then, taking into account that

$$
\nabla V(x) = \nabla(\text{volume}(D_x)) = \int_{\partial D_x} \frac{\gamma_1^1(x,y)\,ds}{|\gamma_2^1(x,y)|},
$$

one can check that the local time term is equivalent to the drift $\frac{1}{2}\nabla(\log V(x))$ as $\epsilon\downarrow 0$.

Remark. One can "guess" to which process the component X_t^ϵ converges, using the following arguments: First, the asymptotic flattening of the boundary ∂D_x in the directions parallel to $\mathbf{R}^d$ allows to expect that the limiting process has a unit diffusion matrix. Second, the invariant measure for X_t^ϵ is the projection of the invariant measure for $(X_t^\epsilon, Y_t^\epsilon)$ on $\mathbf{R}^d$. Since the uniform distribution is invariant for $(X_t^\epsilon, Y_t^\epsilon)$ in D^ϵ, the invariant distribution for X_t^ϵ has the (non-normalized) density $p(x) = \text{volume}(D_x) = V(x)$. On the other hand, the process governed by

$$
\frac{1}{2}\Delta + \nabla\mathcal{V}(x)\cdot\nabla
$$

has an invariant density $\hat{p}(x) = \exp\{2\mathcal{V}(x)\}$. This can be proved easily by checking that $\hat{p}(x)$ is the solution of the stationary forward Kolmogorov equation. Comparing $p(x)$ and $\hat{p}(x)$, we obtain that $\mathcal{V}(x) = \frac{1}{2}\ln V(x)$.

Finally, in this section, we consider shortly the reaction-diffusion equations in narrow branching tubes. Let G^ϵ be the same as in the beginning of this section (see Fig. 16). Consider the problem

$$
\frac{\partial u_l^\epsilon(t,x)}{\partial t} = \frac{D_l}{2}\Delta u_l^\epsilon + f_l(x; u_1^\epsilon, \ldots, u_n^\epsilon), \quad x \in G^\epsilon \subset \mathbf{R}^r,
$$

$$
\frac{\partial u_l^\epsilon(t,x)}{\partial n}\bigg|_{t>0, x\in\partial G^\epsilon} = 0, \quad u_l^\epsilon(0,x) = g_l(x), \quad l \in \{1,\ldots,n\}.
\tag{7.15}
$$

Here the functions $f_l(x; u_1, \ldots, u_n)$ are smooth with bounded first derivatives, $g_l(x)$ are bounded and uniformly continuous. Problems of this type arise, for example, in the theory of nerve impulse propagation and in the combustion theory.

One can expect that the solution of problem (7.15), as in the case of a single linear equation, can be approximated by the solution of the corresponding problem on the graph Γ:

$$
\begin{aligned}
\frac{\partial \bar{u}_l(t, y)}{\partial t} &= \frac{D_l}{2} \frac{\partial^2 \bar{u}_l}{\partial y^2} + f_l(y; \bar{u}_1, \ldots, \bar{u}_n), \quad y \in \Gamma \backslash \{O_1, \ldots, O_m\}, \\
\bar{u}_l(0, y) &= g_l(y), \quad l = 1, \ldots, n,
\end{aligned} \tag{7.16}
$$

supplemented with some gluing conditions at the vertices.

Theorem 7.2 *Let $Y(x)$, $x \in G^\epsilon$, be the point of the graph Γ closest to x (if the closest point is not unique take any of them). Then for any $T \geq 0$,*

$$
\max_{\substack{0 \leq t \leq T, x \in G^\epsilon, \\ l \in \{1, \ldots, n\}}} |u_l^\epsilon(t, x) - \bar{u}_l(t, Y(x))| \to 0 \text{ as } \epsilon \downarrow 0,
$$

where $\bar{u}_1(t, y), \ldots, \bar{u}_n(t, y)$ is the solution of problem (7.16) satisfying the gluing conditions:

$$
\sum_{i: I_i \sim O_k} r_i^{r-1} \frac{\partial \bar{u}_l}{\partial y^i}(O_k) = 0,
$$
$$
\text{if } \lambda_k(\epsilon) \ll \epsilon^{\frac{r-1}{r}} \text{ as } \epsilon \downarrow 0; \ l = 1, \ldots, n;
$$

$$
\sum_{i: I_i \sim O_k} r_i^{r-1} \frac{\partial \bar{u}_l}{\partial y^i}(O_k) + \frac{\varrho_k^r \Gamma(\frac{1}{2}) \Gamma(\frac{r+1}{2})}{\Gamma(\frac{r+2}{2})} \frac{\partial^2 \bar{u}_l}{\partial y^2}(O_k) = 0, \tag{7.17}
$$
$$
\text{if } \lim_{\epsilon \downarrow 0} \lambda_k(\epsilon) \epsilon^{-\frac{r-1}{r}} = 1; \ l = 1, \ldots, n;
$$

$$
\frac{d\bar{u}_l(t, O_k)}{dt} = f_l(O_k, \bar{u}_1(t, O_k), \ldots, \bar{u}_n(t, O_k)),
$$
$$
\text{if } \lambda_k(\epsilon) \gg \epsilon^{\frac{r-1}{r}} \text{ as } \epsilon \downarrow 0.
$$

The problem (7.16)-(7.17) has a unique solution.

To prove this theorem, one can consider the Markov processes $(X_t^\epsilon, P_x^\epsilon)$ in $G^\epsilon \cup \partial G^\epsilon$ governed by the operator $\frac{D_l}{2} \Delta$ inside G^ϵ and having normal reflection on ∂G^ϵ, $l = 1, \ldots, n$. Then the solution of (7.15) satisfies the equations

$$
\begin{aligned}
u_l^\epsilon(t, x) = E_x^{\epsilon, l} g_l(X_t^{\epsilon, l}) + \int_0^t E_x^{\epsilon, l} f_l(X_t^{\epsilon, l}, u^\epsilon(t - s, X_s^{\epsilon, l})) \, ds, \\
t \geq 0, \ x \in G^\epsilon \cup \partial G^\epsilon, \ l = 1, \ldots, n.
\end{aligned} \tag{7.18}
$$

Since the functions f_l are Lipschitz continuous, system (7.18) has a unique solution. This solution can be constructed by successive approximations.

Now let Y_t^l, $l = 1, \ldots, n$ be the process on Γ governed by the operators $\frac{D_l}{2} \frac{d^2}{dy^2}$ and the gluing conditions (7.17) (the third of conditions (7.17) should be replaced by: $\bar{u}_l''(O_k) = 0$ if $\lambda_k(\epsilon) \gg \epsilon^{(r-1)/r}$). Then the solution of problem (7.16)-(7.17) satisfies the equations

$$\bar{u}_l(t, y) = E_y g_l(Y_t^l) + \int_0^t E_y f_l(Y_s^l, \bar{u}(t - s, Y_s^l)) \, ds,$$
$$y \in \Gamma, \quad t \geq 0, \quad l = 1, \ldots, n. \tag{7.19}$$

The problem (7.19) has a unique solution which can be considered as a generalized solution of (7.16)-(7.17). Under certain mild assumptions, this generalized solution is smooth and is the classic one. We will not specify those conditions here but will deal with the generalized solution.

Using the convergence of the processes $Y(X_t^{\epsilon,l}) = Y_t^{\epsilon,l}$ to the process Y_t^l and some a priori bounds of the continuity module of the functions $u_l^\epsilon(t, x)$, one can prove the statement of Theorem 7.2.

Remark. Note that, at the points O_k with $\lambda_k(\epsilon) \gg \epsilon^{(r-1)/r}$ as $\epsilon \downarrow 0$, the evolution of the functions $\bar{u}_l(t, O_k)$ is independent of $\bar{u}(t, y)$ outside such a vertex O_k. This evolution is described by the system of ordinary differntial equations given by the gluing condition at O_k.

8   Wave Fronts in Reaction-Diffusion Equations

A. N. Kolmogorov, I. G. Petrovskii and N. S. Piskunov considered in 1937 [KPP] the following problem:

$$\frac{\partial u(t,x)}{\partial t} = \frac{D}{2}\frac{\partial^2 u}{\partial x^2} + f(u), \ t > 0, x \in R^1$$

$$u(0,x) = \chi^-(x) = \begin{cases} 1, & x \le 0 \\ 0, & x > 0. \end{cases}$$

(8.1)

Here $D > 0$, $f(u) = c(u)\cdot u$; the function $c(u)$ is supposed to be Lipschitz continuous, positive for $u < 1$ and negative for $u > 1$, and such that $c = c(0) = \max_{0 \le u \le 1} c(u)$. Denote by $\mathcal{F}_1$ the class of such functions $f(u)$. It is easy to check that $u(t,x)$ for each $t > 0$ is a strictly monotone function decreasing from 1 as $x \to -\infty$ to 0 as $x \to \infty$. Thus there exists a unique $m = m(t)$ such that $u(t, m(t)) = \frac{1}{2}$. It was proved in [KPP] that $\lim_{t\to\infty} \frac{1}{t}m(t) = \sqrt{2cD}$, and that $u(t, m(t) + z) \to \vartheta(z)$ as $t \to \infty$, where $\vartheta(z)$, $-\infty < z < \infty$ is the solution of the problem

$$\frac{D}{2}\vartheta''(z) + \alpha\vartheta' + f(\vartheta(z)) = 0, \ -\infty < z < \infty,$$

$$\lim_{z\to\infty} \vartheta(z) = 0, \ \lim_{z\to-\infty} \vartheta(z) = 1, \ \vartheta(0) = \frac{1}{2}$$

(8.2)

for $\alpha = \sqrt{2cD}$. Problem (8.2) is solvable for any $\alpha \ge \sqrt{2cD}$, and the solution is unique. Roughly speaking this means that the solution of (1.1) behaves for large t as a running wave $\vartheta(x - \alpha t)$. It can be characterized by its shape $\vartheta(z)$ and by the speed $\alpha = \sqrt{2cD}$.

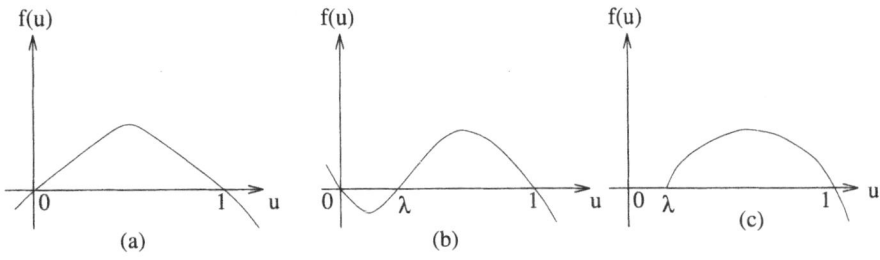

(a) (b) (c)

Figure 20

Other forms of the nonlinear term in (8.1) are also of interest (Fig 20). We say that $f(u), u \in R^1$, belongs to the class $\mathcal{F}_2$, if $f(0) = f(\lambda) = f(1) = 0, 0 < \lambda < 1, f(u) > 0$ for $u < 0$ and for $\lambda > u > 1$, and $f(u) < 0$ for positive $u \notin [\lambda, 1]$ (Fig 20 b). In this case, the "local" dynamical system $\dot{u} = f(u), u \in R^1$, has two stable equilibrium points at $u = 0$, and at $u = 1$ separated by the unstable rest point at $u = \lambda$. Because of this, the nonlinear terms $f(\cdot) \in \mathcal{F}_2$ are called bistable.

The solution of problem (8.1) with a bistable nonlinearity behaves for large t, as a running wave $v(x - \alpha t)$ as well. The asymptotic speed α is defined by the diffusion coefficient D and the function $f \in \mathcal{F}_2$ in a unique way. But, in general, one cannot give a simple formula for the asymptotic speed as was done in the KPP case. The asymptotic shape $v(z)$, again, is the solution of problem (8.2). This problem for $f \in \mathcal{F}_2$ is solvable only for one α (which is the asymptotic speed for problem (8.1)). We consider here, mostly, the KPP case. The bistable case will be discussed shortly in Sec. 9. The results of Sec. 10 are applicable to different types of the nonlinear term.

The problem considered in [KPP] arose in connection with a biological problem: expansion of an advantaged gene. At the same time, a similar problem was considered in [Fi]. Later such kinds of equations appeared in combustion theory, in chemical kinetics, in biophysics. Systems of such equations, so called reaction-diffusion systems (RDE's), describe chemical reaction combined with spatial diffusion of the particles. If n substances take part in the process, then the evolution of their densities $u_k(t, x)$ is described by the RDE-system:

$$
\begin{cases}
\dfrac{\partial u_k(t, x)}{\partial t} & = \quad L_k u_k + f_k(x, u_1, \ldots, u_n), \\[2mm]
u_k(0, x) & = \quad g_k(x), \quad x \in \mathbf{R}^r, \ t > 0, \ k = 1, \ldots, n
\end{cases}
\tag{8.3}
$$

Here L_k is an elliptic second order operator

$$
L_k = \frac{1}{2} \sum_{i,j=1}^{r} a_k^{ij}(x) \frac{\partial^2}{\partial x^i \, \partial x^j} + \sum_{i=1}^{r} b_k^i(x) \frac{\partial}{\partial x^i},
$$

governing the motion of the particles of k-th type; the functions $f_k(x, u_1, \ldots, u_n)$ describe the chemical reaction. If the problem is considered in a bounded volume, some boundary conditions should be added to (8.3).

RDE-systems, on one hand, are relatively simple. The equations are close to linear equations; they are linear in the derivatives. Different equations in (8.3) are connected just through the terms independent of the derivatives. On the other hand, the solutions of RDE-systems can have very rich dynamics. Even when the equations and initial conditions are invariant with respect to all spatial shifts, the system (8.3) still is a dynamical system in $\mathbf{R}^n$ of a general form.

The most attractive problems for RDE's concern the formation of different kinds of space-time organized structures, like running waves, space-time periodic solutions, stable equilibrium solutions having less symmetry than the equations, etc. A new term even appeared for the problems concerning the formation of such patterns – Synergetics. All these features, together with a large number of applications, made RDE's an attractive subject for mathematicians.

The results proved in [KPP] are the first of this type. We will consider different kinds of generalizations of that problem.

As already mentioned, the asymptotic behavior of the solution of problem (8.1) has two characteristics: the shape and the speed of the wave. One can introduce the asymptotic speed, which is the most important characteristic, independently of the shape. The number α^* is called the asymptotic speed as $t \to \infty$ for problem (1.1) if for any $h > 0$

$$\lim_{t \to \infty} \sup_{x > (\alpha^* + h)t} u(t, x) = 0, \quad \lim_{t \to \infty} \inf_{x < (\alpha^* - h)t} u(t, x) = 1$$

It follows from [KPP] that such an α^* exists and is equal to $\sqrt{2cD}$. We will now explain why this statement is a corollary of the asymptotics of the probabilities of large deviations for the Wiener process. Using the Feynman-Kac formula, one can write the equation

$$u(t, x) = E\chi^- \left(x + \sqrt{D}W_t\right) \exp\left\{\int_0^t c\left(u(t - s, x + \sqrt{D}W_s)\right) ds\right\}; \quad (8.4)$$

here W_t is the one-dimensional Wiener process starting at zero. Since $f(\cdot) \in \mathcal{F}_1$, $c(u) \le c(0) = c$, and we derive from (8.4) that

$$0 \le u(t, x) \le E\chi^- \left(x + \sqrt{D}W_t\right) e^{ct} = e^{ct} P\left\{\sqrt{D}W_t \le -x\right\}. \quad (8.5)$$

Let $x = \alpha t$, $\alpha > 0$. Then the event $\left\{W_t \le \frac{-\alpha}{\sqrt{D}}t\right\}$ is a large deviation for the Wiener process W_t as $t \to \infty$, and

$$\log P\left\{W_t \le \frac{-\alpha}{\sqrt{D}}t\right\} \sim -\frac{\alpha^2 t}{2D} \quad \text{as } t \to \infty. \quad (8.6)$$

We obtain from (8.5) and (8.6):

$$0 \le u(t, \alpha t) \le e^{ct} P\left\{W_t < -\frac{\alpha t}{\sqrt{D}}\right\} \asymp \exp\left\{t\left(c - \frac{\alpha^2}{2D}\right)\right\}. \quad (8.7)$$

The bound (8.7) allows us to conclude that for any $h > 0$

$$\lim_{t \to \infty} \sup_{x > t(\sqrt{2cD} + h)} u(t, x) = 0 \quad (8.8)$$

and thus $\alpha^* \le \sqrt{2cD}$.

To prove the lower bound for α^* we will use (8.8) and the large deviation estimates in the functional space for the one-dimensional Wiener process.

Consider the broken line $ABCF$ (Fig. 21):
$$A = (t\sqrt{2cD}, t), \ B = (t\sqrt{2cD}, t(1 - \mu)), \ C = (2t\mu\sqrt{2cD}, \mu t), \ F = (-\mu t, 0):$$

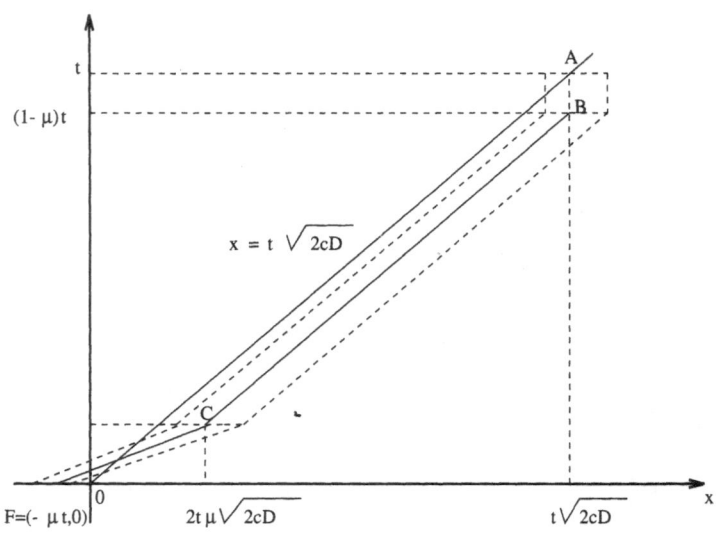

Figure 21

μ is a small positive parameter which is choosen later. Note that CB has the same slope $\sqrt{2cD}$ as $A0$. Let $\kappa = \frac{1}{2}\left(\mu \wedge \mu\sqrt{2cD}\right)$. Define $\pi(s)$ as the x-coordinate of the line $ABCF$ with t-coordinate s, and

$$M = M_\kappa = \{(s,x) : 0 \le s \le t, |x - \pi(s)| < \kappa t\}.$$

It is easy to see that for $\kappa = \frac{1}{2}(\mu \wedge \mu\sqrt{2cD})$ the intersection $M_\kappa \cap \{t = 0\}$ consists of an interval situated to the left from the origin on the t-axis.

Denote by $\mathbf{1}_M$ the indicator function of the set

$$\left\{\phi \in C_{0T} : \max_{0 \le s \le t} |\phi_s - \pi(t - s)| < \kappa t\right\} \subset C_{0t}$$

Let $\tilde{\pi}(s) = \frac{1}{t}\pi(st)$, $0 \le s \le 1$. The function $\tilde{\pi}(s)$ is independent of t. Using the self-similarity property of the Wiener process, one can write :

$$E\mathbf{1}_M\left(t\sqrt{2cD} + \sqrt{D}W.\right) = P\left\{\max_{0 \le s \le t}\left|t\sqrt{2cD} + \sqrt{D}W_s - \pi(t - s)\right| < \kappa t\right\}$$

$$= P\left\{\max_{0 \le s \le t}\left|\sqrt{2cD} + \frac{\sqrt{D}}{\sqrt{t}}W_{s/t} - \frac{1}{t}\pi\left(t\left(1 - \frac{s}{t}\right)\right)\right| < \kappa\right\}$$

$$= P\left\{\max_{0 \le s \le 1}\left|\sqrt{2cD} + \frac{\sqrt{D}}{\sqrt{t}}W_s - \tilde{\pi}(1 - s)\right| < \kappa\right\}.$$

$$\tag{8.9}$$

From the large deviation bounds for ϵW_s, $0 \leq s \leq 1$, in C_{01}, we derive that the last probability is greater than or equal to

$$\exp\left\{-\frac{t}{2D}\left[\int_0^1 |\dot{\pi}(s)|^2 \, ds + \beta\right]\right\}$$

for any $\beta > 0$, if t is large enough. Now, taking into account that BC has the slope $\sqrt{2cD}$, we conclude

$$\int_0^1 |\dot{\pi}(s)|^2 \, ds = \int_\mu^{1-\mu} |\dot{\pi}(s)|^2 \, ds + A(\mu) = (1 - 2\mu) \cdot 2cD + A(\mu).$$

One can easily check that $A(\mu) \to 0$ as $\mu \downarrow 0$. Equality (8.9) and the last bounds result in the following inequality: for any h_1 one can choose $\mu > 0$ so small that for t large enough

$$E\mathbf{1}_M\left(t\sqrt{2cD} + \sqrt{D}W.\right) \geq \exp\left\{-t(c + h_1)\right\}. \tag{8.10}$$

Now, note that if $\mathbf{1}_M(\phi.) = 1$, then, according to (8.8), $\max_{\mu t \leq s \leq (1-\mu)t} u(t - s, \phi_s) \to 0$ as $t \to \infty$. Using the boundedness and the Lipschitz continuity of $c(u)$, $0 \leq u \leq 1$, we derive that, if $\mathbf{1}_M\left(t\sqrt{2cD} + \sqrt{D}W.\right) = 1$, then for any $h_2 > 0$

$$\int_0^t c\left(u(t - s, t\sqrt{2cD} + W_s)\right) \, ds \geq \int_{\mu t}^{(1-\mu)t} c\left(u(t - s, t\sqrt{2cD} + W_s)\right) \, ds$$

$$\geq t(1 - 2\mu)(c - h_2) \tag{8.11}$$

for t large enough. It follows from (8.4),(8.10) and (8.11) that

$$u(t, t\sqrt{2cD}) \geq E\mathbf{1}_M\left(t\sqrt{2cD} + W.\right)\exp\left\{\int_0^t c\left(u(t - s, t\sqrt{2cD} + W_s)\right) ds\right\}$$

$$\geq \exp\{(c - h_2)(t - 2\mu)\}\exp\{-t(c + h_1)\} \geq \exp\{-2t(h_1 + h_2)\}$$

for t large enough. Thus

$$\lim_{t \to \infty} \frac{1}{t} \ln u(t, t\sqrt{2cD}) = 0. \tag{8.12}$$

Let us prove now that for any $h > 0$

$$\lim_{t \to \infty} \inf_{x < t(\sqrt{2cD} - h)} u(t, x) \geq 1 \tag{8.13}$$

Assume that this is wrong: there exists $\lambda > 0$ and arbitrary large t and $x^* = x^*(t) < t\left(\sqrt{2cD} - h\right)$ such that $u(t, x^*) < 1 - \lambda$. Then one can consider a domain

$D = D_t = \left\{ (s,x) : x < s\sqrt{2cD},\ s \geq 0,\ u(s,x) < 1 - \frac{\lambda}{2} \right\}$, containing (t,x^*). Let $\hat{\tau} = \min\{s : (\sqrt{D}W_s, t - s) \notin D\}$. The boundary ∂D consists of two parts: the part belonging to the line $x = s\sqrt{2cD}$, $0 \leq s \leq t$, and the rest of ∂D (Fig. 22). Denote the first part by ∂D_1 and the second by ∂D_2.

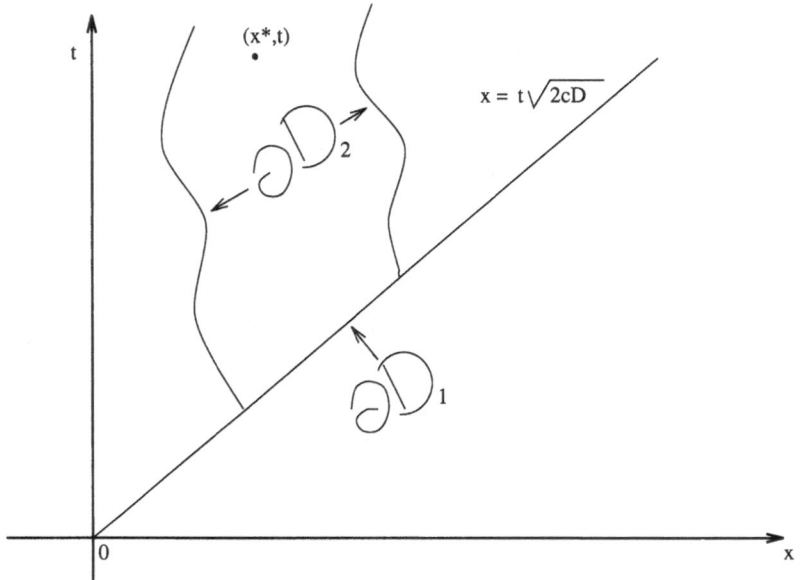

Figure 22

It is easy to check that

$$P_{x^*} \left\{ \hat{\tau} < \frac{ht}{4},\ \left(t - \hat{\tau}, \sqrt{D}W_{\hat{\tau}} \right) \in \partial D_1 \right\} \to 0 \tag{8.14}$$

as $t \to \infty$. Using (8.4) and the strong Markov property, one can write:

$$u(t,x^*) = E_{x^*} u\left(t - \hat{\tau}, \sqrt{D}W_{\hat{\tau}} \right) \exp\left\{ \int_0^{\hat{\tau}} c\left(u(t-s, \sqrt{D}W_s) \right) ds \right\}. \tag{8.15}$$

Using (8.12), (8.14), and taking into account that $c(u(t-s,x)) > d = d(\lambda) > 0$ on ∂D_2, we obtain from (8.15) the bound:

$$u(t,x^*) \geq \left(1 - \frac{\lambda}{2} \right) \cdot P_{x^*} \left\{ \left(t - \hat{\tau}, \sqrt{D}W_{\hat{\tau}} \right) \in \partial D_2 \right\}$$

$$+ P_{x^*} \left\{ \left(t - \hat{\tau}, \sqrt{D}W_{\hat{\tau}} \right) \in \partial D_1,\ \hat{\tau} > \frac{ht}{4} \right\} \cdot \exp\left\{ \frac{hdt}{4} \right\} > 1 - \frac{3\lambda}{4}$$

for t large enough. The last inequality contradicts our assumption that $u(t, x^*) <$ $1 - \lambda$, and thus (8.13) holds. If we observe that, because of the maximum principle, $u(t, x) \leq 1$, then it follows from (8.8) and (8.13) that $\alpha^* = \sqrt{2cD}$ is the asymptotic speed in problem (8.1).

The notion of asymptotic speed can be introduced in a more general situation, and the large deviation approach allows us to calculate it.

Let G be a bounded domain in R^r with a smooth boundary. Consider the following problem in the tube $R^1 \times G$ (Fig 23):

$$\frac{\partial u(t, x, y)}{\partial t} = \frac{D}{2} \Delta_{x,y} u - b \frac{\partial u}{\partial x} + f(u), \ \ t > 0, \ x \in R^1, \ y \in G,$$

$$\left. \frac{\partial u(t, x, y)}{\partial n(y)} \right|_{t<0, x \in R^1, y \in \partial G} = 0, \ \ u(0, x, y) = \chi^-(x). \tag{8.16}$$

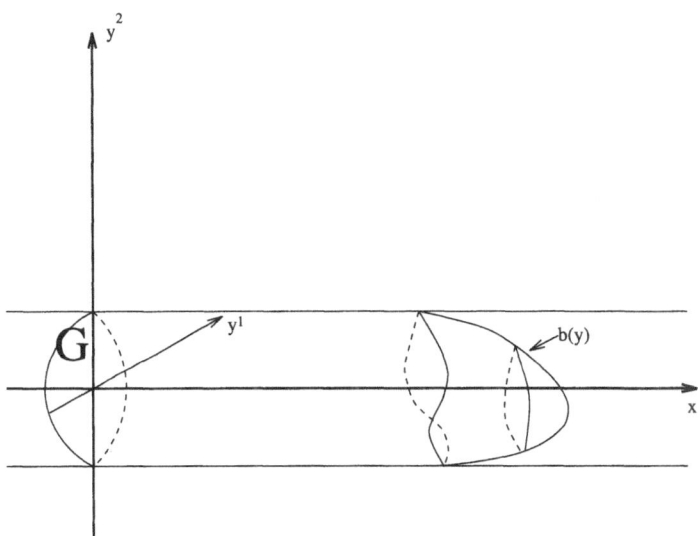

Figure 23

Here $n(y)$ is the outward normal to ∂G at $y \in \partial G$, $\Delta_{x,y}$ is the Laplacian in x and y, $D > 0$, $f(\cdot) \in \mathcal{F}_1$.

We say that α^* is the asymptotic speed for the problem (8.16) as $t \to \infty$, if for any $h > 0$

$$\lim_{t \to \infty} \sup_{\substack{x > (\alpha^* + h)t \\ y \in G \cup \partial G}} u(t, x, y) = 0, \quad \lim_{t \to \infty} \inf_{\substack{x < (\alpha^* - h)t \\ y \in G \cup \partial G}} u(t, x, y) = 1.$$

Equation (8.16) describes the evolution of particles which move with diffusivity D in a flow having a velocity b directed along the x-axis, and take part in a 'chemical reaction' governed by the nonlinear term $f(u)$.

Note, that equation (8.16) is the backward Kolmogorov equation for our process. So that the "physical" drift has the opposite sign. Therefore, (8.16) corresponds to the flow of velocity b.

Of course, one cannot expect that some asymptotic speed will be established if D or b depend on x arbitrarily. Let $D = const$ and let b be independent of x. It follows from KPP results, that if $b = const$ then the asymptotic speed for problem (8.16) is equal to $\alpha^* = b + \sqrt{2cD}$. Now let the velocity of the flow depend on the point of the cross-section: $b = b(y)$, where $b(y)$ is a continuously differentiable function. In the linear case, when $f(u) \equiv 0$, one can check that the asymptotic speed α^* exists and is equal to

$$\bar{b} = \frac{1}{|G|} \int_G b(y)\, dy,$$

where $|G|$ is the volume of $G \in \mathbb{R}^r$. The last statement is a result of averaging in the y-variables: the uniform distribution is invariant for the diffusion process in $G \cup \partial G$ governed inside by the operator $\frac{D}{2}\Delta_y$ and with the normal reflection on ∂G.

If $b(y) \not\equiv const$ and $f \in \mathcal{F}_1$, one might expect that, because of the same averaging, the asymptotic speed will be equal to $\bar{b} + \sqrt{2cD}$. But it turns out that this is not the case. The real asymptotic speed α^* is larger than $\bar{b} + \sqrt{2cD}$ if $b(y) \neq \bar{b}$. Besides the average speed of the flow $\bar{b}$ and the term $\sqrt{2cD}$, which is the result of the interaction between the diffusion in x-direction and the nonlinear term, α^* will have one more positive summand, which is caused by the interplay between the nonlinear term, the deviations of $b(y)$ from $\bar{b}$ and the diffusion in the y-direction.

For the sake of brevity let $D = 1$. Consider the eigenvalue problem

$$\frac{1}{2}\Delta_y \phi_\alpha(y) + \alpha \tilde{b}(y)\phi_\alpha(y) = \lambda(\alpha)\phi_\alpha(y) y \in G, \quad \left.\frac{\partial \phi_\alpha(y)}{\partial n(y)}\right|_{\partial G} = 0. \tag{8.17}$$

Here $\alpha \in \mathbb{R}^1$ is a parameter, $\tilde{b}(y) = b(y) - \bar{b}$. Let $\lambda = \lambda(\alpha)$ be the eigenvalue corresponding to the positive eigenfunction. As is well known, such an eigenfunction exists; $\lambda(\alpha)$ is simple, real and larger than the real part of all other eigenvalues. The function $\lambda(\alpha)$ is continuously differentiable. One can prove that, if Y_t is the Wiener processs in G with the normal reflection on ∂G, then the following equality holds:

$$\lambda(\alpha) = \lim_{t\to\infty} \frac{1}{t} \ln E_y \exp\left\{\alpha \int_0^t \tilde{b}(y_s)\, ds\right\}. \tag{8.18}$$

Using the Hölder inequality and (8.18) we conclude that $\lambda(\alpha)$ is convex:

$$
\begin{aligned}
\lambda(p\alpha + q\beta) &= \lim_{t\to\infty} \frac{1}{t} \ln E_y \exp\left\{(\alpha p + \beta q)\int_0^t \tilde{b}(y_s)\,ds\right\} \\
&= \lim_t \frac{1}{t} \ln E_y \left[\exp\left\{\alpha\int_0^t \tilde{b}(y_s)\,ds\right\}\right]^p \left[\exp\left\{\beta\int_0^t \tilde{b}(y_s)\,ds\right\}\right]^q \\
&\le p\lim_{t\to\infty}\frac{1}{t}\ln E_y \exp\left\{\alpha\int_0^t \tilde{b}(y_s)\,ds\right\} \\
&\quad + q\lim_{t\to\infty}\frac{1}{t}\ln E_y \exp\left\{\beta\int_0^t \tilde{b}(y_s)\,ds\right\} \\
&= p\lambda(\alpha) + q\lambda(\beta), \quad p+q=1; \; p,q>0; \; \alpha,\beta\in\mathbf{R}^1.
\end{aligned}
$$

It is easy to check that $\lambda(\alpha) \ge 0$, and $\lambda(\alpha) = 0$ only for $\alpha = 0$. If $B_+ = \max_{G\cup\partial G}\tilde{b}(y)$ and $B_- = \min G\cup\partial G\tilde{b}(y)$, then $\lambda(\alpha) \le B_+\alpha$ for $\alpha \ge 0$, and $\lambda(\alpha) \le B_-\alpha$ for $\alpha \le 0$.

Consider the Legendre transform $L(\beta)$ of $\lambda(\alpha)$: $\quad L(\beta) = \sup_\alpha [\alpha\beta - \lambda(\alpha)]$.

Taking into account the properties of $\lambda(\alpha)$, one can see that $L(\beta)$ is a convex function, finite in a neighborhood of the origin and equal to $+\infty$ for $\beta \notin [B_-, B_+]$, $L(0) = \max_\alpha(-\lambda(\alpha)) = 0$, and $L(\beta) > -\lambda(0) = 0$ for $\beta \ne 0$.

Denote by $H(\alpha)$ the Legendre transform of $L(\beta) + \frac{\beta^2}{2}$: $H(\alpha) = \sup_\beta \left[\alpha\beta - \frac{\beta^2}{2} - L(\beta)\right] = \frac{\alpha^2}{2} - \inf_\beta\left[\frac{(\alpha-\beta)^2}{2} + L(\beta)\right]$. One can check that the function $\frac{\alpha^2}{2} - H(\alpha)$ is the Legendre transformation of $\frac{\beta^2}{2} + \lambda(\beta)$. For any $c > 0$ consider the equation

$$\frac{\alpha^2}{2} - H(\alpha) = c. \tag{8.19}$$

One can derive from the properties of the function $H(\alpha)$, that equation (8.19) has a unique positive solution for each positive c. This solution is larger than $\sqrt{2c}$. Thus for each continuously differentiable function $\tilde{b}(y)$, $y \in G\cup\partial G$, $\int_G \tilde{b}(y)\,dy = 0$, and any $c > 0$ we defined a number $\alpha^* = \alpha^*[\tilde{b}(\cdot), c]$, which is the positive root of the equation (8.19).

Now we are in a position to write down an expression for the asymptotic speed for problem (8.16)

Theorem 8.1 *Let $u(t, x, y)$ be the solution of problem (8.16), $f \in \mathcal{F}_1$ and $c = \left.\frac{df(u)}{du}\right|_{u=0}$. Then the asymptotic speed $\hat{\alpha}$ is equal to*

$$\hat{\alpha} = \bar{b} + D\alpha^*\left[D^{-1}\tilde{b}(\cdot), \frac{c}{D}\right], \quad where \quad \bar{b} = \frac{1}{|G|}\int_G b(y)\,dy, \; \tilde{b}(y) = b(y) - \bar{b},$$

and the function $\alpha^*[\cdot, \cdot]$ is defined above.

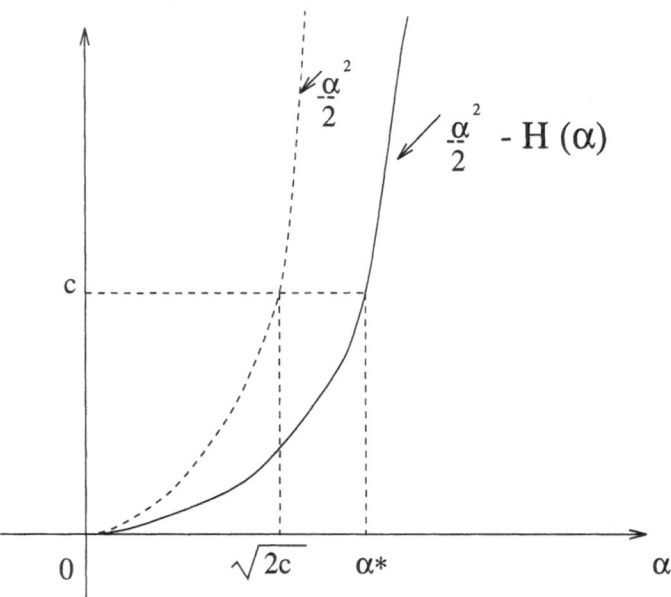

Figure 24

We will outline *the proof* assuming that $D = 1$, $\bar{b} = 0$. The general case can be easily reduced to this one. Consider the Markov process (X_t, Y_t) in $\mathrm{R}^1 \times \{G \cup \partial G\}$, where Y_t is the standard Wiener process in G with the instantaneous normal reflection on the boundary, and X_t is defined by the equation

$$X_t = x - \int_0^t b(Y_s)\, ds + W_t, \qquad (8.20)$$

where W_t is one-dimensional Wiener process independent of Y.

Using the Feynman-Kac formula, one can write down the following equation for the solution of problem (8.16):

$$u(t, x, y) \;=\; E_{x,y}\chi^-(X_t)\exp\left\{\int_0^t c\left(u(t-s, X_s, Y_s)\right)ds\right\}$$
$$c(u) \;=\; u^{-1}f(u), \; t \geq 0, \; x \in \mathrm{R}^1, \; y \in G \cup \partial G.$$

Since $f(\cdot) \in \mathcal{F}_1$, $\max_{0 \leq u} c(u) = c(0) = c$, and one can derive:

$$0 \leq u(t, x, y) \leq E_{x,y}\chi^-(X_t)e^{ct} = e^{ct}P_{x,y}\{X_t < 0\}. \qquad (8.21)$$

To estimate the probability in (8.21) for large t and $x \sim zt$, $z > 0$, introduce the action function for the deviations of order 1 of $\xi_t = \frac{1}{t}\int_0^t b(y_s)\, ds$ and $\frac{1}{t}W_t$ from their mean values as $t \to \infty$.

As it is shown in [FW3], Ch.7, for any $z_1 < z_2$

$$P_y\left\{\xi_t \in [z_1, z_2]\right\} \asymp \exp\left\{-t\min_{z_1 \le z \le z_2} L(z)\right\}, \quad t \to \infty,$$

where the sign '$\asymp$' means the logarithmic equivalence, and $L(z)$ is the Legendre transform of the first eigenvalue $\lambda(\alpha)$ of problem (8.17).

The large deviations for the Gaussian random variable $\frac{1}{t}W_t$ have the asymptotics:

$$P\left\{\frac{1}{t}W_t \in [z_1, z_2]\right\} \asymp \exp\left\{-\frac{t}{2}\cdot\min_{z_1 \le z \le z_2} z^2\right\}, \quad t \to \infty.$$

Since ξ_t and W_t are independent we conclude that the action function for the pair $(\frac{1}{t}W_t, \xi_t)$ is equal to $t(\frac{z^2}{2} + L(z_2))$. According to the contraction principle ([FW1], Theorem 3.3.1) the action function for $\xi_t + \frac{1}{t}W_t$ is equal to

$$t\cdot\min_{z_1}\left\{\frac{(z - z_1)^2}{2} + L(z_1)\right\} = \frac{z^2}{2} - H(z),$$

where $H(z)$ was introduced above. Using this action function we can calculate the logarithmic asymptotic of the probability in (8.21) when $x = zt$ and $t \to \infty$:

$$\lim_{t\to\infty} t\ln P_{zt,y}\{X_t < 0\} = -\left[\frac{z^2}{2} - H(z)\right].$$

From (8.21) and the last equality we obtain:

$$0 \le u(t, x, y) \le \exp\left\{t\left[c - \left[\frac{z^2}{2} - H(z)\right] + o_t(1)\right]\right\}. \tag{8.22}$$

Bound (8.22) shows that

$$\lim_{t\to\infty}\sup_{\substack{x > (\alpha^* + h)t \\ z \in G \cup \partial G}} u(t, x, y) = 0$$

if $h > 0$ and α^* is the positive root of the equation (8.19).

To prove the second equality in the definition of the asymptotic speed one needs the action functional in the space of trajectories. Then one can use a construction similar to one used above, when we proved the lower bound for the asymptotic speed for the KPP-equation (See [F10]). We omit this prove.

To demonstrate more explicitly the effect of increase of the asymptotic speed due to deviations of $b(y)$ from $\bar{b}$, consider the case $G = (-1, 1) \in R^1$ and assume that $D = 1$, $b(y) = \bar{b} + \delta\tilde{b}(y) + o(\delta)$, $\delta \downarrow 0$. Then one can determine [F10] that the asymptotic speed in the problem (8 16) is equal to

$$\hat{\alpha} = \bar{b} + \sqrt{2c} + 2\delta^2\sqrt{2c}\int_{-1}^{1}\left(\int_{-1}^{x}\tilde{b}(z)\,dz\right)^2 dx + o(\delta^2), \quad \delta \downarrow 0.$$

Here $\bar{b}$ is the average speed of the flow, $\sqrt{2c}$ is the KPP speed, and the rest is due to the deviation $b(y)$ from $\bar{b}$.

One can consider some generalizations of the problem (8.16). Let, for example, not only the drift coefficient b depend on y, but the diffusivity and the nonlinear term as well:

$$\frac{\partial u(t,x,y)}{\partial t} = \frac{D(y)}{2}\Delta_{x,y}u - b(y)\frac{\partial u}{\partial x} + f(y,u),$$

$$t > 0, \ x \in \mathbb{R}^1, \ y \in G, \ \left.\frac{\partial u}{\partial n}\right|_{t>0,y\in\partial G,x\in\mathbb{R}^1} = 0, \ u(0,x,y) = \chi^-(x). \tag{8.23}$$

We assume that the coefficients are smooth enough and $f(y,\cdot) \in \mathcal{F}_1$ for each $y \in G\cup\partial G$, $f(y,u) = c(y,u)\cdot u$, $c(y) = c(y,0)$. To calculate the asymptotic speed one should consider the eigenvalue problem:

$$\frac{D(y)}{2}\Delta_y\phi + \left(\alpha_1 D(y) + \alpha_2 b(y) + \alpha_3 c(y)\right)\phi(y)$$

$$= \lambda(\alpha_1,\alpha_2,\alpha_3)\phi(y), \ y \in G, \ \left.\frac{\partial\phi}{\partial n}\right|_{\partial G} = 0. \tag{8.24}$$

Let $\lambda(\alpha_1,\alpha_2,\alpha_3) = \lambda(\alpha)$ be the eigenvalue corresponding to the positive eigenfunction, and let $L(\beta_1,\beta_2,\beta_3) = L(\beta)$ be the Legendre transform of $\lambda(\alpha)$. The function $L(\beta)$ as well as $\lambda(\alpha)$ is convex. Consider the equation

$$\sup_{z_1,z_2,z_3}\left[z_3 - \frac{(v-z_2)^2}{2z_1} - L(z_1,z_2,z_3)\right] = 0. \tag{8.25}$$

One can prove that (8.24) has a unique root ϑ^* such that

$$\bar{b} = \left(\int_G \frac{dy}{D(y)}\right)^{-1}\int_G \frac{b(y)}{D(y)}\,dy \le \vartheta^* < \infty.$$

The constant ϑ^* is the asymptotic speed for problem (8.23) [F10].

Let, for example, $b = 0$, $D = const > 0$, and $f(y,\cdot) \in \mathcal{F}_1$ for each $y \in G\cup\partial G$. Then if we define λ as the first eigenvalue of the problem

$$\frac{D}{2}\Delta\phi(y) + c(y)\phi(y) = \lambda\phi(y),$$

$$y \in G, \ \left.\frac{\partial\phi(y)}{\partial n(y)}\right|_{\partial G} = 0,$$

the asymptotic speed will be equal to $\vartheta^* = \sqrt{2D\lambda}$. Note that if $c(y) = c = const$ than $\lambda = c$ and we have the KPP result. If $c(y) \not\equiv c$ then $\lambda > \bar{c} = \frac{1}{|G|}\int_G c(y)\,dy$.

Consider now a linear parabolic equation in the cylinder $\{t > 0, x \in R^1, y \in G\}$ with nonlinear boundary condition:

$$
\begin{aligned}
\frac{\partial u(t,x,y)}{\partial t} &= \tfrac{1}{2}\Delta u(t,x,y), \ t > 0, \ x \in R^1, \ y \in G, \\
u(0,x,y) &= \chi^-(x), \ \frac{\partial u(t,x,y)}{\partial t} + f(y,u) \mid_{t>0, x\in R^1, y\in\partial G} = 0.
\end{aligned}
\tag{8.26}
$$

Roughly speaking, it means that the "chemical reaction" occurs just on the boundary of the tube $R^1 \times (G \cup \partial G)$. An asymptotic speed will be established in this case as well.

As before, we assume that $f(y,\cdot) \in \mathcal{F}_1$ for each $y \in \partial G$. Then $f(y,u) = c(y,u) \cdot u, c(y,0) = c(y) = max_{u\geq 0}c(y,u)$.

Denote by $Z_t = (X_t, Y_t)$ the Wiener process in the tube $R^1 \times (G \cup \partial G)$ with normal reflection on the boundary. Let L_t be the local time for the process Z_t on the boundary. The process (Z_t, L_t) satisfies the following equation

$$
dZ_t = dW_t + \chi_{\partial G}(Z_t)n(Z_t)dL_t, \ Z_0 = (x,y), \ L_0 = 0.
$$

Here W_t is the Wiener process in R^{r+1}, $\chi_{\partial G}(z)$ is the indicator function of ∂G; $n(z)$, $z \in \partial G$, is the unit inward normal to ∂G. This stochastic equation defines (Z_t, L_t) in a unique way (See, for example, [F6],§1.6).

Using the generalized Feynman-Kac formula, one can write the following equation for the solution of (8.26):

$$
u(t,x,y) = E_{x,y}\chi^-(x)(X_t)\exp\{\int_0^t c(Y_s, u(t-s, X_s, Y_s)dL_s\}. \tag{8.27}
$$

Taking into account the fact that the components X_t, Y_t are independent, and that $c(y,u) \leq c(y)$, we derive from (8.27)

$$
0 \leq u(t,x,y) \leq P_x\{X_t \leq 0\} \cdot E_y \exp\{\int_0^t c(Y_s)dL_s\}. \tag{8.28}
$$

We have for the one-dimensional Wiener process X_t:

$$
\lim_{t\to\infty} t^{-1} \log P_{\alpha t}\{X_t \leq 0\} = -\frac{\alpha t}{2}, \alpha > 0. \tag{8.29}
$$

Consider the eigenvalue problem

$$
\frac{1}{2}\Delta\phi(y) = \lambda\phi(y), \ y \in G \qquad \frac{\partial\phi(y)}{\partial n(y)} + c(y)\phi(y)\Big|_{y\in\partial G} = 0,
$$

and let λ be the eigenvalue corresponding to the positive eigenfunction. Such an eigenvalue exists and is unique. One can prove that

$$
\lim_{t\to\infty} t^{-1} \log E_y\{\int_0^t c(Y_s)dL_s\} = \lambda. \tag{8.30}
$$

Combining (8.28), (8.29) and (8.30) we conclude that for any $h > 0$

$$\lim_{t \to \infty} \sup_{x > (\sqrt{2\lambda} + h)t, y \in G \cup \partial G} u(t, x, y) = 0.$$

Using the large deviation estimates for the process X_t and $\int_0^t c(Y_s) dL_s$, one can prove that

$$\lim_{t \to \infty} t^{-1} \log u(t, t\sqrt{2\lambda}, y) = 0$$

uniformly in $y \in G \cup \partial G$. Taking into account the fact that $0 \le u(t, x, y) \le 1$, and using the standard arguments (compare with [F6], Ch. 6) we derive that

$$\lim_{t \to \infty} \inf_{x < (\sqrt{2\lambda} - h)t, y \in G \cup \partial G} u(t, x, y) \;\; = \;\; 1, \forall h > 0.$$

Thus the asymptotic speed for the problem (8.26) is equal to $\sqrt{2\lambda}$.

If we consider equation (8.16), but replace the Neumann conditions by the Dirichlet condition $u(t, x, y) = 0$ on the boundary, the wave front propagation will be observed only if $c(y) = c(y, 0)$ is large enough compared with the size of the cross-section (cf. [F10]).

I will mention that if the initial function in (8.16) is different from $\chi^-(x)$, one can observe the wave front propagation with a speed larger than in the case of initial function $\chi^-(x)$. In particular, if, say, the initial function for large x is equivalent to $\exp\{-\beta x\}$, $\beta > 0$, then the asymptotic speed tends to infinity as $\beta \downarrow 0$.

To calculate the asymptotic speed in the equation (8.16), we need the logarithmic asymptotic of the probabilities of large deviations for the process X_t defined by (8.20). In the problems considered in this section Y_t was the Wiener process in G with the normal reflection on ∂G. Of course, the same approach allows to consider the case when Y_t is a general non-degenerate diffusion process in a compact phase space. But one can choose something else. For example, let $G = \{1, \ldots, n\}$ and Y_t be a continuous time Markov chain in G with the intensities of the transitions $C_{ij} > 0$, $i \ne j$:

$$P\{Y_{t+\Delta} = j | Y_t = i\} = C_{ij} \cdot \Delta + o(\Delta), \;\; \Delta \downarrow 0.$$

Then the problem (8.16) should be replaced by a problem for the PDE-system, corresponding to the process (X_t, Y_t), where Y_t is our Markov chain and X_t is defined by (8.20):

$$\frac{\partial u_k(t, x)}{\partial t} = \frac{1}{2} \frac{\partial^2 u_k}{\partial x^2} - b_k \frac{\partial u_k}{\partial x} + \sum_{j=1}^{n} C_{kj} \cdot (u_j - u_k) + f(u_k),$$

$$t > 0, \;\; x \in \mathrm{R}^1, \;\; k = 1, \ldots, n, \;\; u_k(0, x) = g(x). \tag{8.31}$$

One can consider a more general system when the diffusivities and the nonlinear terms are different in the different equations. It would be a counterpart of (8.23). But we have to restrict ourselves, for brevity, to the case (8.31).

Note, that if the equations in (8.31) are disconnected, i.e., all $C_{kj} = 0$, then we have different asymptotic speeds in the different equations: the speed for k-th equation is equal to $b_k + \sqrt{2c}$. But if $C_{kj} > 0$, $k \neq j$, a common asymptotic speed will be established. This speed can be found as follows: Denote by $q = (q_1, \ldots, q_n)$ the stationary distribution of the Markov chain, and let $\bar{b} = \sum_{k=1}^{n} b_k q_k$. Consider the $n \times n$ matrix $C(\alpha) = \tilde{C}_{ij}(\alpha))$ with $\tilde{C}_{ij}(\alpha) = C_{ij}$ for $i \neq j$, and $\tilde{C}_{ii}(\alpha) = -\sum_{\substack{j=1 \\ i \neq j}}^{n} C_{ij} + \alpha b_i$. Let $\lambda = \lambda(\alpha)$ be the eigenvalue of $C(\alpha)$ corresponding to the eigenvector with positive coordinates; $\lambda(\alpha)$ will be a smooth convex function. Let

$$\mathcal{A}(\beta) = \sup_{\alpha} \left[\alpha\beta - \frac{\alpha^2}{2} - \lambda(\alpha) \right].$$

Then the common asymptotic speed for system (8.31) is equal to the positive root of the equation $\mathcal{A}(\vartheta) = c = f'(0)$. The proof is similar to the proof of Theorem 8.1.

One can consider a system with different diffusion coefficients and the non-linear terms (but of $\mathcal{F}_1$-class) in different equations:

$$\begin{cases} \dfrac{\partial u_k(t,x)}{\partial t} = \dfrac{D_k}{2} \dfrac{\partial^2 u_k}{\partial x} + f_k(u_k) + \sum_{j=1}^{n} C_{kj} \cdot (u_j - u_k), \quad t > 0, \ x \in \mathbb{R}^1, \\ u_k(0,x) = g_k(x), \quad k = 1, \ldots, n. \end{cases}$$

If all $C_{kj} = 0$, in general, different asymptotic speeds will be established in different equations. If $C_{kj} > 0$ a common speed will be established here as well. This common speed can be larger than the speeds of the separated equations. This is a result from the fact that 'the particles' may use one type for multiplication and another for displacement. Such a "convexization effect" appears in many problems concerning wave front propagation for KPF-type equations (See [F10]).

Finally, in this section, I will mention one more effect. So far we have considered isotropic diffusion. Now we examine wave front propagation in the tube, when the diffusion coefficients along the tube and across it have a different order. Suppose that the diffusivity in the y-variables is much smaller than in the x-direction:

$$\frac{\partial u(t,x,y)}{\partial t} = \frac{D(y)}{2} \frac{\partial^2 u}{\partial x^2} + \frac{\epsilon}{2} \Delta_y u + f(y,u), \quad t > 0, \ x \in \mathbb{R}^1, \tag{8.32}$$

$$u(0,x,y) = \chi^-(x), \ \frac{\partial u(t,x,y)}{\partial n} \Big|_{t>0, x \in \mathbb{R}^1, y \in \partial G} = 0$$

Here $f(y, \cdot) \in \mathcal{F}_1$ for $y \in G \cup \partial G, 0 < \epsilon \ll 1$.

If $t \to \infty$ but $\epsilon t \to 0$ there is not enough time for mixing in the y-variables, and therefore, roughly speaking, an asymptotic speed as $t \to \infty$ will be established for each $y \in G \cup \partial G$ separately. If $\epsilon \tilde{t}^{-1}$, one can expect that a common asymptotic speed for all $y \in G \cup \partial G$ will be established. To consider this case, let us rescale space and time. Let $u^\epsilon(t, x, y) = u(t/\epsilon, x/\epsilon, y)$, where $u(t, x, y)$ is the solution of (8.32). Then $u^\epsilon(t, x, y)$ is the solution of the following problem:

$$\frac{\partial u^\epsilon(t, x, y)}{\partial t} = \frac{\epsilon D(y)}{2} \frac{\partial^2 u^\epsilon}{\partial x^2} + \frac{1}{2} \Delta_y u^\epsilon + \frac{1}{\epsilon} f(y, u^\epsilon), \ t > 0, \ x \in R^1, \quad (8.33)$$

$$u^\epsilon(0, x, y) = \chi^-(x), \quad \frac{\partial u^\epsilon(t, x, y)}{\partial n}\Big|_{t>0, x \in R^1, y \in \partial G} = 0.$$

We are interested in the behavior of $u^\epsilon(t, x, y)$ as $\epsilon \downarrow 0$. As before, let Y_t be the Wiener process in $G \cup \partial G$ with normal reflection on the boundary, and

$$X_t^\epsilon = x + W(\epsilon \int_0^t D(Y_s)ds),$$

where $W(t)$ is the Wiener process in R^1 independent of $Y.$. The Feynman-Kac formula implies:

$$u^\epsilon(t, x, y) = E_{x,y}\chi^-(X_t) \exp\{\frac{1}{\epsilon} \int_0^t c(Y_s, u^\epsilon(t - s, X_s^\epsilon, Y_s))ds\}.$$

Again, since $f(y, \cdot) \in \mathcal{F}_1$, we can conclude that

$$0 \leq u^\epsilon(t, x, y) \leq$$
$$E_y[\exp\{\frac{1}{\epsilon} \int_0^t c(Y_s)ds\} \cdot P\{x + W(\epsilon \int_0^t D(Y_s)ds) < 0 \mid Y_s, 0 \leq s \leq t\}]. \quad (8.34)$$

Since $W.$ and $Y.$ are independent random processes, the conditional probability on (8.34) can be calculated explicitely. We need only the logarithmic asymptotic of this probability as $\epsilon \downarrow 0$:

$$\lim_{\epsilon \downarrow 0} \epsilon \log P\{x + W(\epsilon \int_0^t D(Y_s)ds) < 0 \mid Y_s, 0 \leq s \leq t\} = -x^2(2 \int_0^t D(Y_s)ds)^{-1}.$$

From this, one can derive that the right hand side of (8.34) is logarithmically equivalent to

$$E_y \exp\{\frac{1}{\epsilon}[\frac{t}{t} \frac{1}{t} \int_0^t c(Y_s)ds - \frac{(x/t)^2}{\frac{2}{t} \int_0^t D(Y_s)ds}]\}. \quad (8.35)$$

Denote by $\mu_t(\Gamma), \Gamma \subset G \cup \partial G$, the occupation time for the process Y_s in the time interval $[0, t)$:

$$\mu_t(\Gamma) = \int_0^t \chi_\Gamma(Y_s)ds,$$

where $\chi_\Gamma(\cdot)$ is the indicator function of the measurable set $\Gamma \subset G \cup \partial G$. Then (8.35) can be rewritten as follows.

$$E_y \exp\{\frac{t}{\epsilon}[\int_{G \cup \partial G} c(z)\mu_t(dz) - (\frac{x}{t})^2 (2\int_{G \cup \partial G} D(z)\mu_t(dz))^{-1}]\} \qquad (8.36)$$

Since the support of the distribution of the process $Y_s, 0 \le s \le t, Y_0 = y$, is the whole space of continuous functions on $[0,t]$ with values in $G \cup \partial G$, starting at $y \in G \cup \partial G$, it is easy to check that the expectation (8.36) is logarithmically equivalent as $\epsilon \downarrow 0$ to

$$\exp\{\frac{t}{\epsilon} \sup_{\nu:\nu(G \cup \partial G)=1} [\int_{G \cup \partial G} c(z)\nu(dz) - (\frac{x}{t})^2 (2\int_{G \cup \partial G} D(z)\nu(dz))^{-1}]\}, \qquad (8.37)$$

where the supremum is taken over all measures ν on $G \cup \partial G$ such that $\nu(G \cup \partial G)=1$.
  To calculate the supremum in (8.37), consider the mapping $T : G \cup \partial G \to R^2$:

$$y \to T(y) = (c(y), D(y)),$$

and denote by $\mathcal{A}$ the image of $G \cup \partial G$ in R^2:

$$\mathcal{A} = \{z \in R^2 : z = T(y), y \in G \cup \partial G\},$$

It is clear that $\mathcal{A}$ is a compact connected set in R^2. Let $\mathcal{A}_{conv}$ be the convex hull of the set $\mathcal{A}$. Then it is easy to check that the supremum in (8.37) is equal to

$$\max_{(z_1,z_2) \in \mathcal{A}_{conv}} [z_1 - \frac{x^2}{2t^2 z_2}] = \gamma(\frac{x}{t}). \qquad (8.38)$$

The function $\gamma(\alpha)$ is strictly decreasing for $\alpha > 0$, and therefore the equatioon $\gamma(\alpha) = 0$ has a unique positive root $\hat{\alpha}$. One can check that

$$\hat{\alpha} = \max_{(z_1,z_2) \in \mathcal{A}_{conv}} \sqrt{2z_1 z_2}. \qquad (8.39)$$

The function $\gamma(\alpha)$ is negative when $\alpha > \hat{\alpha}$ and positive for $\alpha < \hat{\alpha}$. Combining (8.34)–(8.38) together, we conclude that

$$\lim_{\epsilon \downarrow 0} u^\epsilon(t, x, y) = 0$$

uniformly in any compact subset of $\{(t, x, y) : t > 0, y > \hat{\alpha}t, y \in G \cup \partial G\}$. One can show (See [F10]), that

$$\lim_{\epsilon \downarrow 0} u^\epsilon(t, x, y) = 1$$

uniformly in any compact subset of $\{(t, x, y) : t > 0, x < \hat{\alpha}t, y \in G \cup \partial G\}$. Thus we can conclude that $\hat{\alpha}$ is the asymptotic speed for problem (8.33). Two cases are possible: $\hat{\alpha} = \max_{y \in G \cup \partial G} \sqrt{2c(y)D(y)}$ or $\hat{\alpha} > \max_{y \in G \cup \partial G} \sqrt{2c(y)D(y)}$. In the first case, there exists $y^* \in G \cup \partial G$ such that $\hat{\alpha} = \sqrt{2c(y^*)D(y^*)}$. This means that the speed $\hat{\alpha}$ is equal to the asymptotic speed which will be established along the line $\{(x, y^*) : x \in R^1\}$. The speed along any other line $\{(x, y) : y \in G \cup \partial G, x \in R^1\}$ in the absense of mixing in the y-direction is less than or equal to the speed obtained when $y = y^*$. The mixing helps to establish the maximal speed for all $y \in G \cup \partial G$. In the second case, a common speed $\hat{\alpha}$ will be established which is bigger than the asymptotic speed for each fixed $y \in G \cup \partial G$. In this case, roughly speaking, the particles defining motion of the front use some points of the cross-section $G \cup \partial G$ for multiplication (for chemical reaction) and other points of $G \cup \partial G$ for the motion in the x-direction.

9   Wave Fronts in Slowly Changing Media

We consider in this section one more generalization of the KPP result. Our goal
here is to study wave fronts in non-homogeneous media. Of course, one cannot
expect the existence of an asymptotic speed without any assumption about the
coefficients: they must be, in an asymptotic sense, homogeneous. For example, one
can prove the existence of an asymptotic speed if the coefficients and $f(x, u)$ are
periodic in x or close to periodic (see [F6]). If the coefficients and $f(x, u)$ are space-
homogeneous random fields with some ergodic properties, one also can expect that
an asymptotic speed will be established. The last problem is, actually, solved just
in the one-dimensional case (see [F6] and references there).

There exists another way to study the non-homogeneous problems. Assume
that the coefficients of the operator L and the non-linear term are changing slowly:

$$\frac{\partial u(t, x)}{\partial t} = \frac{1}{2} \sum_{i,j=1}^{r} a^{ij}(\epsilon x) \frac{\partial^2 u}{\partial x^i \, \partial x^j} + f(\epsilon x, u),$$

$$t > 0, \ x \in \mathbb{R}^r, \ 0 < \epsilon \ll 1.$$

Then we can expect that the asymptotic speed in a proper time scale will be
established locally near each point of the phase space, and the motion of the front
on the whole will be governed by those local speeds.

Assume that the coefficients $a^{ij}(x)$ and functions $f(x, y)$ are Lipschitz continuous, the matrix $(a^{ij}(x)) = a(x)$ is positively defined, uniformly in $x \in \mathbb{R}^r$, and
$f(x, \cdot) \in \mathcal{F}_1$ for each $x \in \mathbb{R}^r$.

Rescale space and time: Set $u^\epsilon(t, x) = u(t/\epsilon, x/\epsilon)$. Then $u^\epsilon(t, x)$ satisfies the
equation

$$\frac{\partial u^\epsilon(t, x)}{\partial t} = \frac{\epsilon}{2} \sum_{i,j=1}^{r} a^{ij}(x) \frac{\partial^2 u^\epsilon}{\partial x^i \, \partial x^j} + \frac{1}{\epsilon} f(x, u^\epsilon), \quad t > 0, \ x \in \mathbb{R}^r. \qquad (9.1)$$

Let us add to this equation an initial condition

$$u^\epsilon(0, x) = g(x) \geq 0, \qquad (9.2)$$

where $g(x)$ is a continuous bounded function. We also allow $g(x)$ to have simple
discontinuities, but such that the closure $[G_0]$ of the support $G_0 = \{x \in \mathbb{R}^r :
g(x) > 0\}$ is equal to the closure of its interior (G_0). A typical example of such a
function is the indicator function of a set $G_0 \subset \mathbb{R}^r$ with $[G_0] = [(G_0)]$.

It is easy to prove, that the solution $u^\epsilon(t, x)$ of problem (9.1)-(9.2) exists and
is unique, and

$$0 \leq u^\epsilon(t, x) \leq \left(1 \vee \sup_{x \in \mathbb{R}^r} g(x)\right), \quad \overline{\lim}_{\epsilon \downarrow 0} u^\epsilon(t, x) \leq 1. \qquad (9.3)$$

These inequalities follow from the maximum principle for linear parabolic equations and from the negativity of $c(x,u) = u^{-1}f(x,u)$ for $u > 1$.

Let X_t^ϵ be the diffusion process in $\mathbf{R}^r$ corresponding to the operator

$$\epsilon L = \frac{\epsilon}{2} \sum_{i,j}^{r} a^{ij}(x) \frac{\partial^2}{\partial x^i \, \partial x^j}.$$

Then the Feynman-Kac formula gives the following equation for $u^\epsilon(t,x)$:

$$u^\epsilon(t,x) = E_x g\left(X_t^\epsilon\right) \exp\left\{\frac{1}{\epsilon} \int_0^t c\left(X_s^\epsilon, u^\epsilon(t-s, X_s^\epsilon)\right) ds\right\}, \tag{9.4}$$
$$t \geq 0, \quad x \in \mathbf{R}^r, \quad f(x,u) = c(x,u) \cdot u.$$

Since $f(x, \cdot) \in \mathcal{F}_1$ for each $x \in \mathbf{R}^r$, $c(x) = c(x,0) \geq c(x,u)$, and we derive from (9.4):

$$0 \leq u^\epsilon(t,x) \leq E_x g\left(X_t^\epsilon\right) \exp\left\{\frac{1}{\epsilon} \int_0^t c\left(X_s^\epsilon\right) ds\right\}. \tag{9.5}$$

As we have already seen in Section 3, the action functional for the family X_t^ϵ, $0 \leq t \leq T$, in C_{0T} as $\epsilon \downarrow 0$ has the form

$$S_{0T}(\phi) = \begin{cases} \dfrac{1}{2} \displaystyle\int_0^T \sum_{i,j=1}^{r} a_{ij}(\phi_s)\dot{\phi}_s^i \dot{\phi}_s^j \, ds, & \text{if } \phi \text{ is absolutely continuous,} \\[2mm] +\infty, & \text{for the rest of } C_{0T}. \end{cases}$$

Here $(a_{ij}(x)) = (a^{ij}(x))^{-1}$. The normalizing coefficient is equal to ϵ^{-1}, The functional $S_{0T}(\phi)$ is semicontinuous from below.

Using the large deviation estimates we derive from (9.5) that

$$0 \leq u^\epsilon(t,x) \leq E_x g\left(X_t^\epsilon\right) \exp\left\{\frac{1}{\epsilon} \int_0^t c(X_s^\epsilon) \, ds\right\}$$
$$\asymp \exp\left\{\frac{1}{\epsilon} \sup\left[\int_0^t c(\phi_s) \, ds - S_{0T}(\phi) : \phi \in C_{0T}, \phi_0 = x, \phi_t \in [G_0]\right]\right\}, \tag{9.6}$$

where $[G_0]$ is the closure of $G_0 \subset \mathbf{R}^r$. Call to mind that '$\asymp$' is the sign of logarithmic equivalence as $\epsilon \downarrow 0$. Denote by

$$R_{0t}(\phi) = \int_0^t c(\phi_s) \, ds - S_{0T}(\phi),$$
$$V_1(t,x) = \sup\left[R_{0t}(\phi) : \phi_0 = x, \ \phi_t \in [G_0]\right].$$

It is easy to check that $R_{0t}(\phi)$ is upper semicontinuous and $V_1(t,x)$ is a Lipschitz continuous function. It follows from (9.4), that

$$\lim_{\epsilon \downarrow 0} u^\epsilon(t,x) = 0, \quad \text{if } V_1(t,x) < 0. \tag{9.7}$$

The convergence in (9.7) is uniform in (t, x) from any compact subset of $\{t \geq 0, x \in R^r : V_1(t, x) < 0\}$.

If we could show that $\lim_{\epsilon \downarrow 0} u^\epsilon(t, x) = 1$ in the area where $V_1(t, x) > 0$ then the equation $V_1(t, x) = 0$ would give us the position of the interface (wavefront) separating regions where $u^\epsilon(t, x)$ tends to zero and to 1 as $\epsilon \downarrow 0$. As we will see now this is true, but under certain assumptions. Without these assumptions the result is a bit more complicated, and we will formulate it later.

We say that the condition (N) is fulfilled, if for any (t, x) such that $V_1(t, x) = 0$

$$\sup \{R_{0t}(\phi) : \phi \in C_{0t}, \ \phi_0 = x, \ \phi_t \in [G_0], \ V_1(t - s, \phi_s) < 0 \text{ for } 0 < s < t\} = 0.$$

We will see later examples where (N) holds and where it does not hold.

Theorem 9.1 *Let $u^\epsilon(t, x)$ be the solution of problem (9.1)-(9.2) with $f(x, \cdot) \in \mathcal{F}_1, x \in R^r$. Then*

$$\lim_{\epsilon \downarrow 0} u^\epsilon(t, x) = 0$$

uniformly in (t, x) from any compact subset of $\{t \geq 0, x \in R^r : V_1(t, x) < 0\}$. If (N) holds,

$$\lim_{\epsilon \downarrow 0} u^\epsilon(t, x) = 1,$$

uniformly in any compact subset of $\{t \geq 0, x \in R^r : V_1(t, x) > 0\}$, and the equation $V_1(t, x) = 0$ defines the position of the front at time t.

The first statement was proved. We will just outline the proof of the last statement, to explain where condition (N) is used. One can find the detailed proof in [F6] (Section 6.2).

Let $V_1(t, x) = 0$. Then, because of condition (N), the supremum in the definition of $V_1(t, x)$ can be taken not over all functions $\phi_s, \phi_0 = x, \phi_t \in G_0$, but rather over the part of this set consisting of functions $\phi_s, 0 \leq s \leq t$, such that $(t - s, \phi_s) \in \{(\tau, y) : V_1(\tau, y) < 0\} = \xi_-$ for $0 < s < t$. We already know that $\lim_{\epsilon \downarrow 0} u^\epsilon(t, x) = 0$ for $(t, x) \in \xi_-$. Therefore, for such functions ϕ_s and any $\delta > 0$

$$\left| \int_0^t c(\phi_s, u^\epsilon(t - s, \phi_s)) ds - \int_0^t c(\phi_s) ds \right| < \delta,$$

if ϵ is small enough. This together with the large deviation lower bound for process $X_s^\epsilon, 0 \leq s \leq t$, implies that

$$\lim_{\epsilon \downarrow 0} \epsilon \log u^\epsilon(t, x) = \lim_{\epsilon \downarrow 0} \epsilon \log E_x g(X_t^\epsilon) \exp\{\frac{1}{\epsilon} \int_0^t c(X_s^\epsilon, u^\epsilon(t - s, X_s^\epsilon)) ds\}$$

$$\geq \lim_{\epsilon \downarrow 0} \epsilon \log E_x g(X_t^\epsilon) \exp\{\frac{1}{\epsilon} \int_0^t c(X_s^\epsilon) ds\} = V_1(t, x) = 0.$$

Since $0 \leq u^\epsilon(t,x) \leq (1 \vee \sup_x g(x)) < \infty$, we conclude that $\lim_{\epsilon \downarrow 0} \epsilon \log u^\epsilon(t,x) = 0$ if $V_1(t,x) = 0$. Using the strong Markov property of the process (X_t^ϵ, P_x) and the positivity of $c(x,u)$ for $u < 1$, one can derive the last statement of Theorem 9.1 (compare with Section 8).

Note, that if (N) does not hold, the extremals of the variational problem defining $V_1(t,x)$ can go inside the domain where $u^\epsilon(t,x)$ is far from zero. Then $c(x, u^\epsilon(t,x)) < c(x)$, and the function V_1 does not describe the motion of the front (see the example after Theorem 9.2).

As the first example, consider the case when the nonlinear term is independent of $x : f = f(u)$ [F4]. Then the expression for $V_1(t,x)$ can be simplified:

$$
\begin{aligned}
V_1(t,x) &= ct - \inf\{S_{0t}(\phi) : \phi \in C_{0t}, \ \phi_0 = x, \ \phi_t \in [G_0]\} \\
&= ct - \inf\left\{\int_0^t \sum_{i,j=1}^r a_{ij}(\phi_s)\dot{\phi}_s^i \dot{\phi}_s^j \, ds; \ \phi_0 = x, \ \phi_t \in [G_0]\right\}.
\end{aligned}
\tag{9.8}
$$

Introduce the Riemannian metric $\rho(\cdot,\cdot)$ in R^r corresponding to the form

$$
\sum_{i,j=1}^r a_{ij}(x) \, dx^i \, dx^j.
$$

One can prove that the infimum in the right side of (9.8) is equal to $\frac{1}{2t}\rho^2(x, G_0)$, and

$$
V_1(t,x) = ct - \frac{\rho^2(x, G_0)}{2t}.
$$

One can also check that the condition (N) is fulfilled in this example [F4]. Thus we conclude from Theorem 9.1 that the 'excited' area at time t (the area where $u^\epsilon(t,x)$ is close to 1) consists of the points of R^r located not farther than $t\sqrt{2c}$ from the support of the initial function. It means that the excited area is growing according to the Huygens principle, and the velocity field is homogeneous and isotropic if calculated in the Riemannian metric $\rho(\cdot,\cdot)$, the absolute value of the velocity field is equal to $\sqrt{2c}$ [F4].

Recall that domains $G_t \subset R^r$, $t \geq 0$, grow according to the Huygens principle with velocity field $v(x,e)$, $x \in R^r$, $e \in R^r$, $|e| = 1$, if

$$
G_{t+h} = \{y \in R^r : \inf_{\phi_0 \in G_t, \, \phi_1 = y} \int_0^1 \frac{\sqrt{\sum_{i,j=1}^r a_{ij}(\phi_s)\dot{\phi}_s^i \dot{\phi}_s^j}}{v(\phi_s, \dot{\phi}_s, |\dot{\phi}_s|)} ds < h\},
$$

for any $t, h \geq 0$. The infimum here is taken over all smooth $\phi_s, 0 \leq s \leq 1$, with values in R^r, connecting the points of G_t and $y \in R^r$. $v(x,e)$ is the speed of the excitation propagation at a point $x \in R^r$ in the direction e.

It is well known that many asymptotic problems for hyperbolic differential equations describing wave processes lead to a Huygens principle. It turns out that the growth of the domain G_t, where the solution of problem (9.1) is close to 1 as $\epsilon \downarrow 0$, also can be described by a Huygens principle.

In particular, if $u^\epsilon(0, x) = g^\epsilon(x)$, where $g^\epsilon(x)$ is positive in a neighborhood of the origin $G_0^\epsilon = \{x \in \mathbb{R}^r : \epsilon^{-1} x \in G_0\}$ and equal to zero outside $[G_0^\epsilon]$, then the excitation at time t will occupy the Riemannian ball $B_t = \{x \in \mathbb{R}^r : \rho(x, 0) < t\sqrt{2c}\}$. Note that the topological structure of B_t can be different from the euclidian ball.

The situation is more complicated if $c(x) \neq const$, even if the condition (N) holds. Consider the following example [F7]: Let $x \in \mathbb{R}^1$, $a^{11}(x) \equiv 1$, $f(x, u) = c(x)u(1 - u)$ where $c(x) = 1$ for $x < 0$ and $c(x) = 1 + x$ for $x > 0$. Let the support of the initial function $G_0 = G_0^a = \{x < a\}$, $a \geq 0$. In this case

$$V_1(t, x) = V_1^a(t, x) = \sup\left\{ \int_0^t \left(1 + \dot{\phi}_s - \frac{\dot{\phi}_s^2}{2} \right) ds; \ \phi_0 = x, \ \phi_t = a \wedge x \right\}.$$

We are interested in $x > a$. Then one can calculate that

$$V_1^a(t, x) = \frac{t^3}{24} + t\left(1 + \frac{a + x}{2}\right) - \frac{(a - x)^2}{2t}$$

By equating $V_1^a(t, x)$ to zero we find the expression for the front position $X_a(t)$ at time t:

$$X_a(t) = \frac{t^2}{2} + a + \sqrt{\frac{t^4}{3} + 2t^2(1 + a)}$$

Taking into account that $X_a(t)$ is strictly increasing, and at each time t its derivative can be represented in the form of a function of $X_a(t)$, it may seem that in this case the front propagation also admits a description with the Huygens principle with an appropriate velocity field. However, the velocity field here turns out to depend on the initial condition. To see this it is sufficient to evaluate $\tilde{a} = X_0(1)$, $X_{\tilde{a}}(1)$ and $X_0(2)$:

$$X_0(1) = \tilde{a} = \frac{1}{2} + \sqrt{\frac{7}{3}},$$

$$X_0(2) = 2 + \sqrt{\frac{40}{3}} \approx 5.6,$$

$$X_{\tilde{a}}(1) = 1 + \sqrt{\frac{7}{3}} + \sqrt{\frac{10}{3}} + \sqrt{\frac{28}{3}} \approx 5.1.$$

If the velocity field did not depend on the initial condition, then the equality $X_0(2) = X_{\tilde{a}}(1)$ would be valid. In our case $X_0(2) > X_{\tilde{a}}(1)$, and thus the velocity field is not of such universal nature as in the previous example.

This example shows also that the motion of the front does not satisfy the Markov property (i.e., the semi-group property) : given the position of the front at time 1, the behavior of the front before $t = 1$ can influence the behavior of the front after time 1.

The evolution of the function $u^\epsilon(t, x)$ satisfies, of course, the semi-group property: if $u^\epsilon(s, x)$ is known for $x \in R^r$, then $u^\epsilon(t, x), t > s$, can be calculated in a unique way. However it turns out, that for $c(x) \neq const$, the evolution of the main term of $u^\epsilon(t, x)$ as $\epsilon \downarrow 0$, which is a step function with values 0 and 1, already loses the semi-group property. To preserve the semi-group property, one must extend the phase space. The phase space should include not just the region where the function $v(t, x) = \lim_{\epsilon \downarrow 0} \epsilon \log u^\epsilon(t, x)$ changes from a negative value to zero (the position of the wave front), but the whole function $v(t, x)$, $x \in R^r$. Then the evolution of $v(t, x)$ will be described by a semigroup (Compare with Remark 9.3 and with Section 10).

In the next example we will show that if $c(x)$ is changing fast enough, the wave front can have jumps. We will, first, choose a discontinuous function as $c(x, 0) = c(x)$ It is done solely for the sake of simplification of the computations. In the final part of the example we will see that the effects are preserved if $c(x)$ is continuous, provided it increases sufficiently fast in some finite interval.

Consider again the one-dimensional problem with $a_{11}(x) \equiv 1$, $c(x, 0) = c_1 > 0$ for $x < h$ and $c(x, 0) = c_2 > 2c_1$ for $x \geq h > 0$. Let $g(x) = \chi^-(x)$, which is the indicator function of $\{x < 0\} \subset R^1$. Inside each of the half-lines $\{x < h\}$ and $\{x > h\}$ Euler's equation for the functional

$$R_{0t}(\phi) = \int_0^t \left(c(\phi_s) - \frac{\dot{\phi}_s^2}{2} \right) ds$$

has the form $\ddot{\phi}_s = 0$. Therefore, the extremals of $R_{0t}(\phi)$ are line segments with vertices on the line $x = h$. This reasoning allows to calculate

$$V(t, x) = \sup \{ R_{0t}(\phi) : \phi_0 = x, \phi_t = 0 \},$$

to check the validity of condition (N) and to describe the wave front motion.

It turns out that up to time $T_0 = (h/c_2)\sqrt{2(c_2 - c_1)}$ the 'excitation' propagates from the point $0 \in R^1$ to the right with the constant velocity $\sqrt{2c_1}$ and reaches the point $T_0\sqrt{2c_1} < h$ at time T_0 (Fig. 25).

On the other hand, simple calculations show that $V(T_0, h) = 0$ and $V(t, h) > 0$ for $t > T_0$. This means that at time T_0 a new source appears at the point $x = h$. From this source a wave propagates in both directions, to the left with velocity $\sqrt{2c_1}$ and to the right with a velocity which is at first larger than $\sqrt{2c_2}$. The waves from the point 0 and from the point h will meet by time $T_1 = (2\sqrt{2c_1})^{-1}(h - T_0\sqrt{2c_1})$. Thus at the time T_0 a jump of the wave front occurs and the excited region has two connected components for $t \in (T_0, T_1)$.

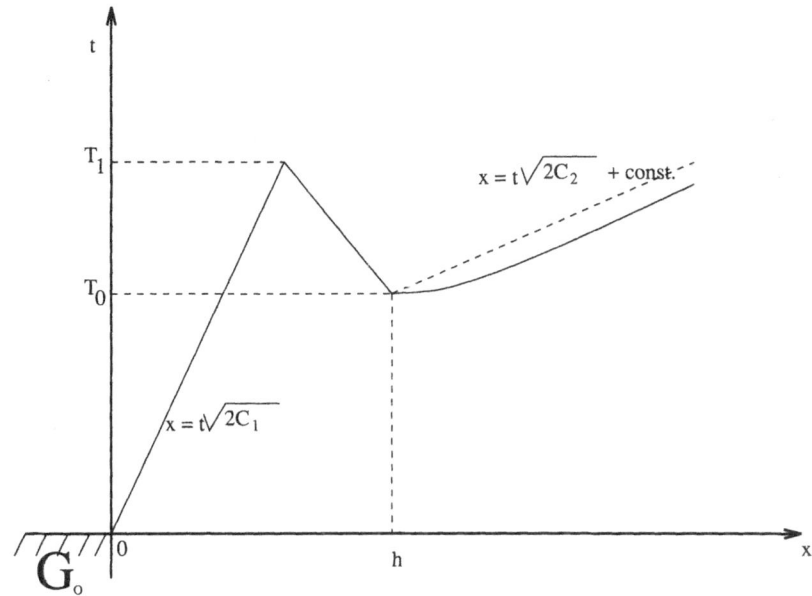

Figure 25

Now, let $c(x,0) = c(x)$ be a smooth monotone function which is equal to c_1
for
$$x < \bar{x} = \frac{1}{2}\left[h + T_0\sqrt{2c_1}\right] + \delta,$$
where δ is small enough, and is equal to c_2 for $x > h$. Using comparision theorems
one can show that for such a function $c(x)$ the excitation reaches the region $\{x >
\bar{x} + \delta\}$ before it reaches the point $\bar{x}$. Thus here we also have a jump of the wave
front.

Thus we have a satisfactory description of the motion of the wave front, provided condition (N) holds. But this condition is not always satisfied. One can show,
for example, that if $r = 1$, $a^{11} \equiv 1$, $G_0 = \{x < 0\}$ and $c(x)$ is a strictly decreasing
function, the condition (N) is not fulfilled. We will give now a description of the
limiting behavior of the solution of problem (9.1)-(9.2) when (N) is not valid. Let
([F8],[FL2])

$$V(t,x) \;=\; \sup\left\{ \min_{0 \le a \le t} \int_0^a \left[c(\phi_s) - \tfrac{1}{2}\sum_{i,j=1}^r a_{ij}(\phi_s)\dot{\phi}_s^i\dot{\phi}_s^j\right] \, ds : \right.$$
$$\left. \phi \in C_{0t} \text{ is absolutely continuous}, \phi_0 = x, \ \dot{\phi}_t \in G_0 \right\}.$$

One can prove that the function $V(t,x)$ is Lipschitz continuous and $V(t,x) \le
0$. If (N) is fulfilled, then $V(t,x) = \min(0, V_-(t,x))$.

Theorem 9.2 *Let $u^\epsilon(t,x)$ be the solution of the problem (9.1)-(9.2), $f(x,\cdot) \in \mathcal{F}_1$ for each $x \in R^r$. Then*

(i) $\lim_{\epsilon \downarrow 0} u^\epsilon(t,x) = 0$ *uniformly in any (t,x) which is contained in a compact subset of $\{(t,x): t > 0, \ x \in R^r, \ V(t,x) < 0\}$;*

(ii) $\lim_{\epsilon \downarrow 0} u^\epsilon(t,x) = 1$ *uniformly in any (t,x) which is contained in a compact subset of the interior of the set $\{(t,x): t > 0, \ x \in R^r, \ V(t,x) = 0\}$.*

The proof of this theorem can be found in [F8], [FL2].

Consider an example [F7] : Let $r = 1$, $a_{11}(x) \equiv 1$, $g(x) = \chi^-(x)$ and $c(x,u)$ be such that $c(x) = c(x,0)$ is a strictly decreasing smooth function for $x \geq 0$. As has been mentioned before, the condition (N) in this case does not hold.

Let $\phi(t)$ be the solution of the equation

$$\dot\psi(t) = \sqrt{2c(\psi(t))}, \ \psi(0) = 0.$$

Because of our assumptions on $c(x)$, $\psi(t)$ increases monotonically, its derivative is positive and decreases as t grows. For any continuous ϕ_s, $\phi_0 = x > 0$, define $\hat{a} = \hat{a}[\phi.] = \min\{s: \phi_s \geq \psi(t-s)\}$. One can prove that

$$V(t,x) = \sup \left\{ \int_0^{\hat{a}[\phi.]} \left[c(\phi_s) - \frac{1}{2}\dot\phi_s^2 \right] ds : \phi \in C_{0t}, \ \phi_0 = x, \ \phi_t = 0 \right\},$$

and that $V(t,x) < 0$ if $x > \psi(t)$ and $V(t,x) = 0$ for $x < \psi(t)$. Thus, according to Theorem 9.2, the wave front at time t is located at the point $\psi(t)$:

$$\lim_{\epsilon \downarrow 0} u^\epsilon(t,x) = \begin{cases} 0, & if \ x > \psi(t), \\ 1, & if \ x < \psi(t). \end{cases}$$

One can describe in this example the propagation of the excitation from the region $\{x < 0\}$ to the right using the Huygens principle with the velocity field $\vartheta(x) = \sqrt{2c(x)}$. One should, however, keep in mind that the front velocity at the point x depends on which direction the wave moves. If at the initial time the region $\{x > 10\}$ were excited (it means that $G_0 = \{x > 10\}$), then the front would propagate to the left (with the same function $c(x)$) with a different velocity and, in particular, in this case the front may have jumps. Note, that the front in the direction where $c(x)$ decreases (for instance, from 0 to the point $x = 10$) is slower than in the direction of increasing $c(x)$ (from $x = 10$ to $x = 0$).

In general, for problems (9.1)–(9.2), the region G_t excited at time t is always greater than or equal to the domain G'_t obtained from G_0 according to the Huygens principle with the velocity field $v(x,e) = \sqrt{2c(x)}$ in the corresponding Riemannian metric. This follows from the maximum principle and the fact that the front propagates according to the Huygens principle with velocity $v(x, e) \equiv \sqrt{2e}$, if $c(x) = c = const$ (See [F6], [F9]). We have seen that G_t can be strictly larger than G_t'. In the last example G_t coincides with G_t' if $G_0 = \{x < 0\}$ and is larger than G_t' if $G_0 = \{x > 10\}$.

We will make now several remarks concerning different generalizations of problem (9.1)-(9.2).

Remark 9.1   Let a number of disjoint connected domains $D_i \subset R^r, i = 0, 1, \ldots, N$, be given. Let the boundaries of the domains be smooth enough. Assume that the function $c(x) = c(x, 0)$ is smooth, positive in $\bigcup_0^N D_i$ and $uc(x, u) = f(x, u) \in \mathcal{F}_1$ for $x \in \bigcup_1^N D_i$, and $c(x) = \max_{u \geq 0} c(x, u) < 0$ for $x \notin [\bigcup_1^N D_i]$. Let the initial function in problem (9.1)-(9.2) be positive for $x \in D_0$ and equal to zero outside $[D_0]$. Will the excitation come to the other 'islands' D_k, $k \neq 0$, and how will this occur?

We say that D_k is switched on at time T, if $\max_{x \in D_k} u^\epsilon(t, x) \to 0$ as $\epsilon \downarrow 0$ for $t < T$ and $\max_{x \in D_k} u^\epsilon(t, x) \to 1$ as $\epsilon \downarrow 0$ for $t > T$.

One can derive from the considerations of this section (see [F9]), that in the generic case there exists a sequence of numbers $i_1, \ldots, i_N \in \{1, \ldots, N\}$ and times $T_1 < T_2 < \ldots < T_N < \infty$ such that D_{i_l} will be switched on at the time T_l, $l = 1, \ldots, N$. These i_l and T_l are defined through the functional $R_{0t}(\phi)$. For example, T_1 and i_1 are defined as follows:

$$T_1 = \min \left\{ t : \max_{1 \leq i \leq N} \sup \{R_{0t}(\varphi) : \phi_0 \in D_0, \phi_t \in D_i\} \right\} = 0;$$

i_1 is the number (unique in the generic case) for which the $\max_{1 \leq i \leq N}$ in the last formula is achieved.

The problem considered in Remark 9.1 arises in a biological model. In [F9] is also mentioned another model, where $c(x, u)$ is replaced by a 'non-local' term. In the last case one can observe propagation not of a wave front but of an impuls of a positive volume. The impuls can move continuously and can have jumps.

Remark 9.2   RDE's describe processes with an interplay between the process of transportation of particles (diffusion) on one hand, and the "chemical reaction" described by the nonlinear term on the other hand. The transportation process must not necessarily be a non-degenerate diffusion. For example, consider the problem

$$\frac{\partial u^\epsilon(t, x, y)}{\partial t} = \frac{1}{2\epsilon} \frac{\partial^2 u^\epsilon}{\partial y^2} + b(x, y) \frac{\partial u^\epsilon}{\partial x} + \frac{1}{\epsilon} f(x, y, u^\epsilon)$$
$$t > 0, \ x \in R^1, \ |y| < 1, \ u^\epsilon(0, x, y) = \chi^-(u), \ \left.\frac{\partial u}{\partial y}\right|_{|y|=1} = 0. \tag{9.9}$$

The y-component of the process $(X_t^\epsilon, Y_t^\epsilon)$ connected with the problem (9.9) is equal to $Y_{t/\epsilon}^1$, where Y_t^1 is the Wiener process with reflection in $[-1, 1]$. The x-component has a finite speed:

$$\dot{X}_t^\epsilon = b\left(X_t^\epsilon, Y_{t/\epsilon}^1\right).$$

Define

$$\bar{b}(x) = \frac{1}{2}\int_{-1}^1 b(x, y)\, dy, \quad \tilde{b}(x, y) = b(x, y) - \bar{b}(x)$$

and consider the eigenvalue problem

$$\frac{1}{2}\phi''(y) + \left(\alpha_1\tilde{b}(x, y) + \alpha_2 c(x, y)\right)\phi(y) = \lambda\phi(y)$$

$$-1 < y < 1, \quad \phi'(\pm 1) = 0.$$

(9.10)

Here $c(x, y) = \lim_{u\downarrow 0} u^{-1}f(x, y, u)$, $f(x, y, \cdot) \in \mathcal{F}_1$; $\alpha_1, \alpha_2 \in \mathbf{R}^1$. Let $\lambda = \lambda(x, \alpha_1, \alpha_2)$ be the first eigenvalue of the problem (9.10). Define $L(x, \beta_1, \beta_2)$ as the Legendre transformation of $\lambda(x, \alpha_1, \alpha_2)$ in α_1, α_2, and

$$V(t, x) = \sup\left\{\phi_t^2 - \int_0^t L\left(\phi_s^1, \dot{\phi}_s^1, \dot{\phi}_s^2\right)\, ds : \phi_0^1 = x, \ \phi_t^1 \le 0, \ \phi_0^2 = 0\right\}.$$

Then under certain assumptions the position of the front at time t can be calculated as the positive root $x^* = x^*(t)$ of the equation $V(t, x^* + \bar{b}t) = 0$. The position, of course, is independent of y.

One can consider as the transportation process Markov processes with jumps.

Remark 9.3   Let $r = 1$, $a^{11}(x) \equiv 1$, $f \in \mathcal{F}_1, c(x) = c = const$ and let $g^\epsilon(x) = 1$ for $x \le 0$ and $g^\epsilon(x) = \exp\{-\frac{\alpha x}{\epsilon}\}$ for $x > 0$. Then one can check that the position X_t^* of the wave front at time t is given as follows

$$X_t^* = \begin{cases} t(c/\alpha + \alpha/2), & for\ \alpha < \sqrt{2c} \\ t\sqrt{2c}, & for\ \alpha \ge \sqrt{2c}. \end{cases}$$

So that the front velocity may be arbitrary large, provided α is small enough. The velocity decreases as α grows, but it is always not less than $\sqrt{2c}$.

Remark 9.4   Consider the mixed problem for the linear heat equation with nonlinear boundary conditions:

$$\frac{\partial u^\epsilon(t, x)}{\partial t} = \frac{\epsilon}{2}\Delta u^\epsilon, \ t > 0, \ x \in \mathbf{R}_+^r = \{x \in \mathbf{R}^r, x^1 > 0\},$$

$$\frac{\partial u^\epsilon(t, x)}{\partial x^1} - \frac{1}{\epsilon}f(x, u^\epsilon)\bigg|_{x^1=0} = 0, \ u^\epsilon(0, x) = g(x) \ge 0$$

Assume that $f(x, \cdot) \in \mathcal{F}_1$, $f(x, u) = c(x, u)u$, $c(x) = c(x, 0)$. Let X_t^ϵ be the diffusion process in $\mathbf{R}_+^r$ governed by the operator $\frac{\epsilon}{2}\Delta$ inside $\mathbf{R}_+^r$ with normal reflection on the boundary. Denote by ξ_s^ϵ the local time on the boundary $x^1 = 0$ associated with the process X_t^ϵ (see, for example. [F6]). Then one can write for $u^\epsilon(t, x)$ an equation:

$$u^\epsilon(t, x) = E_x g(X_t^\epsilon) \exp\left\{\frac{1}{\epsilon} \int_0^t c(X_s^\epsilon, u^\epsilon(t - s, X_s^\epsilon))\, d\xi_s^\epsilon\right\}$$

One can calculate the action functional for the couple $(X_t^\epsilon, \xi_s^\epsilon)$ as $\epsilon \downarrow 0$ and to describe the motion of the front. A number of results and examples concerning the problems with nonlinear boundary conditions one can find in [F6].

Remark 9.5   In these lectures we did not consider nonlinear terms different from KPP type. Let $f(u)$ have a bistable nonlinearity (Fig. 20b): $f(0) = f(\lambda) = f(1) = 0, 0 < \lambda < 1$, $f(u) > 0$ for $u < 0$ and $u \in (\lambda, 1)$, and $f(u) < 0$ for $u > 1$ and $0 < u < \lambda$. The class of such functions $f(u)$ was denoted by $\mathcal{F}_2$. Assume that $f(u)$ is continuously differentiable and $f'(0), f'(\lambda), f'(1)$ are not equal to zero. It is known that the solution of problem (8.1) with such nonlinear term converges to a running wave solution:

$$\mid u(t, x) - v(x - \alpha^* t) \mid \to 0 \text{ as } t \to \infty,$$

where α^* is the unique value of α for which the following problem is solvable:

$$\frac{D}{2} v''(z) + \alpha v'(z) + f(v(z)) = 0, \quad -\infty < z < \infty,$$

$$\lim_{z \to -\infty} v(z) = 1, \quad \lim_{z \to \infty} v(z) = 0.$$

The last problem defines the shape $v(z)$ of the wave as well (See references in [F6]). Therefore, α^* is defined by the nonlinear term $f \in \mathcal{F}_2$ and by the constant $D : \alpha^* = \alpha^*[f, D]$. Note that in the bistable case α^* can be positive or negative: the sign of $\alpha^*[f, D]$ coinsides with the sign of $\int_0^1 f(u)du$. If the sign is positive, the area where $u(t, x) \to 1$ as $t \to \infty$ is growing. If $\int_0^1 f(u)du < 0$, the area where $u(t, x) \to 0$ is expanding.

One can consider the small parameter problem

$$\frac{\partial u^\epsilon(t, x)}{\partial t} = \frac{\epsilon}{2} \sum_{i,j=1}^r a^{ij}(x) \frac{\partial^2 u^\epsilon}{\partial x^i \partial x^j} + \frac{1}{\epsilon} f(x, u^\epsilon), \ t > 0, \ x \in R^r, \quad (9.11)$$

$$u^\epsilon(0, x) = \chi_{G_0}(x), \ f(x, \cdot) \in \mathcal{F}_2.$$

Here $\chi_{G_0}(x)$ is the indicator function of the set $G_0 \subset R^r$, which is assumed to be regular enough.

If $f(x, \cdot) \in \mathcal{F}_2$ for any $x \in R^r$, then $\bar{f}(x, y) = -f(x, 1 - u)$ also belongs to $\mathcal{F}_2$, and the function $v = 1 - u^\epsilon$ satisfies the equation (9.11) with f replaced by

$\bar{f}$. Therefore, it is sufficient in this case to have sufficiently good bounds from one side, say, from below for $u^\epsilon(t, x)$. The upper bound can be obtained as the lower bound for $v^\epsilon(t, x) = 1 - u^\epsilon(t, x)$. Moreover, the negativity of $f(x, u)$ for small u makes jumps of the wave front impossible. Therefore, the propagation of the wave front for $f(x, \cdot) \in \mathcal{F}_2$ always has a local character. The front moves continuously in this case. This allows us to freeze the coefficients in (9.11) and to reduce the problem to the one-dimensional equation (8.1) with $f \in \mathcal{F}_2$. The wave front motion for equation (9.11) with bistable nonlinear term always can be described by the Huygens principle. The corresponding velocity field will be isotropic in the Riemannian metric connected with (9.11). The absolute value of the speed at each point $x \in R^r$ can be expressed through the functional $\alpha^*[f(x, \cdot), D]$ introduced above for the one-dimensional problem and through the diffusion matrix $(a^{ij}(x))$. The exact result can be found in [G] (See also [F6] Ch. 6).

It is worth noting that the probabilistic approach turns out so far less successful in the bistable case than in the KPP case. This is connected partly with the fact that in the bistable case, it is impossible to separate the asymptotic speed and the shape of the wave: in this case the logarithmic asymptotics of

$$u^\epsilon(t, x) = E_x^\epsilon g(X_t^\epsilon) \exp\{\frac{1}{\epsilon} \int_0^t c(X_s^\epsilon, u^\epsilon(t - s, X_s^\epsilon)) ds\}, \quad \epsilon \downarrow 0,$$

is defined by the trajectories going in the transition area where $\lambda < u < 1$, since $c(x, u)$ is negative outside of this interval.

A similar difficulty arises in the case of the nonlinear term shown in Fig 20 c. In the next section we consider an approach which gives some results for the nonlinear terms of various types.

In the conclusion of this section, we will consider systems of reaction-diffusion equations with a small parameter. The qualitative behavior of the solutions of RDE-systems is, of course, much richer than in the case of single equation. Various space-time structured solutions can be observed in the case of systems. This is due to the fact that multi-dimensional dynamical systems defined by nonlinear terms can be much more complicated than one-dimensional system.

We consider here just systems of KPP-type, when, roughly speaking, the dynamical system has one unstable equilibrium point, say, at the origin $u = 0$, and one stable attractor. Then one can expect that the area, where the solution is close to the stable attractor, will expand with time, and this can be described as a motion of an interface (wave front) separating areas where the solution is close to the attractor and where it still has values close to the unstable point.

Consider the following system

$$\frac{\partial u_k^\epsilon(t, x)}{\partial t} = \frac{1}{2} \sum_{i,j=1}^r a_k^{ij}(x) \frac{\partial^2 u_k^\epsilon}{\partial x^i \partial x^j} + \frac{1}{\epsilon} F_k(x, u_1^\epsilon, \dots, u_n^\epsilon),$$

$$t > 0, \ x \in \mathbb{R}^r, \ k = 1, \dots, n.$$

(9.12)

We assume that the coefficients and functions F_k are sufficiently regular and the operators $L_k = \frac{1}{2} \sum_{i,j} a_k^{ij}(x) \frac{\partial^2}{\partial x^i \partial x^j}$, $k = 1, \ldots, n$, are uniformly elliptic. Moreover, we assume that the vector field $u = (u_1, \ldots, u_n) \to F(x, u) = (F_1(x, u), \ldots \ldots, F_n(x, u))$, indexed by $x \in \mathbb{R}^r$, satisfies the following conditions analogous to the KPP conditions in the case of one equation:

A1 There exists $B > 0$ such that for every $x \in \mathbb{R}^r$ the vector field $F(x, u)$ points strictly inward from the boundary of the cube $[0, B]^n$, except at $u = 0$ $F(x, 0) = 0$, and $\inf\{F_k(x, u) : 1 \le k \le n, u \in [0, B]^n, x \in \mathbb{R}^r\} = -M > -\infty$.

A2 Define $C_{km}(x) = \frac{\partial F_k(x, 0)}{\partial u_m}$. Assume that $C_{km}(x)$ are Lipschitz continuous and

$$\sup\{C_{km}(x) : 1 \le k, m \le n, x \in \mathbb{R}^r\} = \bar{\beta} < \infty,$$
$$\inf\{C_{km}(x) : 1 \le k, m \le n, x \in \mathbb{R}^r\} = \beta > 0.$$

Define $C(x) = (C_{km}(x))$. Note that $u = 0$ is an unstable rest point for the field $F(x, u)$ for any $x \in \mathbb{R}^r$.

A3 $F_k(x, u) \le \sum_{m=1}^n C_{km}(x) u_m$, $1 \le k \le n$, $x \in \mathbb{R}^r$, $u \in [0B]^n$. For each $\gamma > 0$ there exists $B' = B'(\gamma) > 0$ (independent of x) such that

$$F_k(x, u) \ge \sum_{m=1}^n (C_{km}(x) - \gamma) u_m, \ 1 \le k \le n, \ x \in \mathbb{R}^r, \ u \in [0, B']^n.$$

An accompanying linear system can be connected with (9.12):

$$\frac{\partial \vartheta_k^\epsilon(t, x)}{\partial t} = \epsilon L_k \vartheta_k^\epsilon(t, x) + \frac{1}{\epsilon} \sum_{i=1}^n C_{ki}(x)(\vartheta_i^\epsilon - \vartheta_k^\epsilon) \tag{9.13}$$
$$t > 0, \ x \in \mathbb{R}^r, \ k = 1, \ldots, n.$$

Let $(X_t^\epsilon, \nu_t^\epsilon)$ be the Markov process in $\mathbb{R}^r \times \{1, \ldots, n\}$ corresponding to the system (9.13). Define

$$Z_t^\epsilon = \left(\int_0^t \chi_1(\nu_s^\epsilon) \, ds, \ldots, \int_0^t \chi_n(\nu_s^\epsilon) \, ds \right),$$

where χ_k is the indicator function of the point $k \in \{1, \ldots, n\}$, so that $\int_0^t \chi_k(\nu_s^\epsilon) \, ds$ is the occupation time at the state k for process ν_s^ϵ during the time interval $[0, t]$.

To study wave front propagation for the system (9.12), one needs the large deviation principle for the family $(X_t^\epsilon, Z_t^\epsilon)$, $0 \le t \le T$, as $\epsilon \downarrow 0$.

The action functional for this family in the uniform topology was found in [FL1]:

$$S_{0T}(\phi, \mu) = \begin{cases} \int_0^T \eta(\phi_s, \dot{\phi}_s, \dot{\mu}_s) \, ds, & \text{if } \phi \text{ and } \mu \text{ are absolutely continuous,} \\ \qquad \sum_{k=1}^n \mu_s^k = s, \ 0 \le s \le T, \\ \qquad \text{and } \mu_s^k \text{ are non-decreasing,} \\ +\infty, & \text{otherwise.} \end{cases}$$

Here $\phi : [0, T] \to \mathbf{R}^r$, $\mu = (\mu^1, \dots, \mu^n) : [0, T] \to \mathbf{R}^n$. The function $\eta(x, q, \beta)$ is the Legendre transformation in p and α of the first eigenvalue $\lambda(x, p, \alpha)$ of the matrix

$$\hat{C}(x) + M(x, p, \alpha),$$

where $\hat{C}(x) = (\hat{C}_{ij}(x))$, $\hat{C}_{ij}(x) = C_{ij}(x)$ for $i \neq j$, $\hat{C}_{kk} = -\sum_{j=1}^n C_{kj}(x)$, and $M(x, p, \alpha)$ is the diagonal matrix with elements

$$M_{kk} = \frac{1}{2} \sum_{i,j=1}^r a_k^{ij}(x) p_i p_j + \alpha_k; \quad k = 1, \dots, n.$$

To describe the law of the wave front motion we need some notations: Let $\zeta(x, q)$ be defined as follows:

$$\zeta(x, q) = \sup_{\beta \in \mathbf{R}^n} \left[\bar{C}(x) \cdot \beta - \eta(x, q, \beta) \right]; \quad x, q \in \mathbf{R}^n,$$

$$\bar{C}(x) = \left(\sum_{m=1}^n C_{1m}(x), \dots, \sum_{m=1}^n C_{nm}(x) \right).$$

Finally, introduce the function

$$V_{G_0}(t, x) = \sup \left\{ \min_{0 \leq a \leq t} \int_0^a \zeta(\phi_s, \dot{\phi}_s) \, ds; \ \phi \in C_{0t}, \ \phi_0 = x, \ \phi_t \in G_0 \right\}.$$

Here G_0 is a domain in $\mathbf{R}^r$. One can prove that $V_{G_0}(t, x)$ is locally Lipschitz continuous; $V_{G_0}(t, x) \leq 0$.

Theorem 9.3 *Consider the Cauchy problem for system (9.13) with initial conditions*

$$u_k^\epsilon(0, x) = g_k(x), \ x \in \mathbf{R}^r; \ k = 1, \dots, n.$$

Assume that the functions $g_k(x)$ are continuous, $0 \leq g_k(x) \leq B$, $k = 1, \dots, n$; $G_0 = \bigcup_{k=1}^n \operatorname{supp} g_k$. Let the conditions (A1), (A2), (A3) hold. Then

(i) *$\lim_{\epsilon \to 0} u_k^\epsilon(t, x) = 0$ for $1 \leq k \leq n$, uniformly in any (t, x) which is contained in a compact subset of $\{(t, x) : t > 0, \ x \in \mathbf{R}, \ V_{G_0}(t, x) < 0\}$;*

(ii) *$\underline{\lim}_{\epsilon \to 0} u_k^\epsilon(t, x) > 0$ for $1 \leq k \leq n$, uniformly in any (t, x) which is contained in a compact subset of the interior of the set $\{(t, x) : t > 0, \ x \in \mathbf{R}^r, V_{G_0}(t, x) = 0\}$.*

The proof of this theorem can be found in [FL2].

I would like to mention some peculiarities of the wave front propagation in the case of systems. For the case of single equation, as we have seen above, if

$c(x) = c = const$ then the wave front always propagates according to the Huygens principle related to the diffusion matrix $(a^{ij}(x))$ and to the constant c. One can show, that for $n \geq 2$ and $C_{ij}(x) = C_{ij} = const$, this is not the case: the propagation of the front can be not necessarily described by a Huygens principle.

In the space homogeneous case, that is when $a_k^{ij}(x) = a_k^{ij}$ and $C_{km}(x) = C_{km}$ are constants,

$$V_{G_0}(t, x) = t \cdot \min_{y \in G_0} \zeta\left(\frac{y - x}{t}\right).$$

This implies that the wave front propagates according to the Huygens principle. But the corresponding velocity field is isotropic and homogeneous in space not in a Riemannian metric, but in a Finsler metric associated with the unit ball $H = \{q \in R : \zeta(q) \geq 0\}$.

Finally, I will mention, so called weakly coupled reaction diffusion equations. Such equations, after proper space-time rescaling, have the form:

$$\frac{\partial u_k^\epsilon(t, x)}{\partial t} = \epsilon L_k u_k^\epsilon + \frac{1}{\epsilon} f_k(x, u_k^\epsilon) + \sum C_{kj}(x)\left(u_j^\epsilon - u_k^\epsilon\right),$$
$$u_k^\epsilon(0, x) = g_k(x) \geq 0, \ t > 0, \ x \in R^r, k = 1, \ldots, n. \tag{9.14}$$

Here $f_k(x, \cdot) \in \mathcal{F}_1$, $C_{kj}(x) > 0$. The law of wave front motion for system (9.14) was described in [F8]. In particular, if $f_k = f_k(u_k)$ are independent of x and if $f_k'(0) = C = const$ is independent of k, the wave front propagates according to the Huygens principle. The corresponding velocity field is isotropic and its absolute value equal to $\sqrt{2C}$ if calculated in the Finsler metric in R^r, generated by the family of unit balls $B(x)$, $x \in R^r$: the ball $B(x)$ at a point x is the convex envelope of the Riemannian unit balls $B_k(x) = \left\{y \in R^r : \sum_{i,j=1}^r a_{ij,k}(x)y^i y^j \leq 1\right\}$, $(a_{ij,k}(x)) = (a_k^{ij}(x))^{-1}, k \in \{1, \ldots, n\}$.

10   Large Scale Approximation for Reaction-Diffusion Equations

We introduce in this section a new assumption concerning the nonlinear term, which allows us to calculate the asymptotics in the bistable case and for other types of local dynamics. But first we give a more general formulation of the problem of interaction between the stochastic transport and multiplication of the particles.

The solution of problem (9.1) satisfies equation (9.4), and we actually study the behavior of the solution of (9.4) as $\epsilon \downarrow 0$. Equation (9.4) can be considered not just in the case where the particle transport is described by a nondegenerate diffusion process, but for other types of the stochastic transport as well.

Let $(Z_t^\epsilon, P_z^\epsilon)$ be a Markov process in a metric space Z, depending on a parameter $\epsilon > 0$. Consider an equation of (9.4) type:

$$u^\epsilon(t, z) \;=\; E_z^\epsilon g(Z_t^\epsilon) \exp\{\frac{1}{\epsilon} \int_0^t c^\epsilon(Z_s^\epsilon, u^\epsilon(t - s, Z_s^\epsilon))ds\} \qquad (10.1)$$

Here $g(z)$ is a bounded and continuous function on Z and $c^\epsilon(z, u)$, $z \in Z$, $u \in R^1$, $\epsilon > 0$, is a bounded Lipschitz continuous function. Denote by $R[u]$ the operator in the right-hand side of (10.1):

$$R[u^\epsilon](t, z) = E_z^\epsilon g(Z_t^\epsilon) \exp\{\frac{1}{\epsilon} \int_0^t c^\epsilon(Z_s^\epsilon, u^\epsilon(t - s, Z_s^\epsilon))ds\}.$$

Since $c(z, u)$ is Lipschitz continuous in u, the operator $R[u]$ is a contraction in the space of continuous bounded functions on $Z \times [0, t]$ provided with the metric $\rho_{0,t}(u, v) = \sup_{0 \le s \le t} |u(s, z) - v(s, z)|$, if t is small enough; ϵ is fixed. This implies that equation (10.1) has a unique continuous solution $u^\epsilon(t, z)$, $t > 0, z \in Z$, for any $\epsilon > 0$. For example, $(Z_t^\epsilon, P_z^\epsilon)$ can be a Markov process in R^r with jumps, or a degenerate diffusion process when the classical PDE-theory has difficulties. A diffusion process in a bounded domain with some boundary conditions can be considered as Z_t^ϵ in (10.1). If $(Z_t^\epsilon, P_z^\epsilon)$ is the Markov process in $R^r \times \{1, \ldots, n\}$ described by (1.16), then (10.1) is connected with systems of reaction-diffusion equations.

Assume now that $\frac{1}{\epsilon} S_{0T}(\phi)$ is the action functional for the family $Z_t^\epsilon, 0 \le t \le T$, as $\epsilon \downarrow 0$, in the space $C_{0T}(Z)$ of continuous functions on $[0, T]$ with values in Z. The space $C_{0T}(Z)$ is provided with the uniform topology. Let, for brevity, $S_{0T}(\phi) = 0$ if and only if $\phi_s \equiv 0$, $0 \le s \le T$.

Consider the KPP case. Let $c^\epsilon(z, u) = c(z, u)$ be independent of ϵ and $f(z, u) = c(z, u)u \in \mathcal{F}_1$ for any $z \in Z$. Set

$$V_1(t, z) = \sup_{\phi \in C_{0t}(Z), \, \phi_0 = x, \phi_t \in G_0} \{\int_0^t c(\phi_s)ds - S_{0t}(\phi)\},$$

where $c(z) = c(z,0)$ and G_0 is the support of the function $g(z)$ which is assumed to be non-negative. Assume that condition (N) is fulfilled: if $V_1(t,z) = 0$, then

$$V_1(t,z) = \sup\Big\{ \int_0^t c(\phi_s)ds - S_{0t}(\phi):$$
$$\phi \in C_{0t}(Z), \phi_0 = z, \phi_t \in G_0, V_1(t-s,\phi_s) < 0 \text{ for } 0 < s < t \Big\}.$$

Then under some mild additional conditions, one can prove that $\lim_{\epsilon \downarrow 0} u^\epsilon(t,z) = 0$ if $V_1(t,z) < 0$, and $\lim_{\epsilon \downarrow 0} u^\epsilon(t,z) = 1$ if $V_1(t,z) > 0$.

If the condition (N) does not hold, we introduce the function $V(t,z)$ (compare with Section 9):

$$V(t,z) = \sup_{\substack{\phi \in C_{0t}(Z) \\ \phi_0 = z, \phi_t \in G_0}} \{ \min_{0 \le a \le t} [\int_0^a c(\phi_s)ds - S_{0a}(\phi)] \}.$$

Then the position of the wave front at a time t is described as the boundary of the set $\{z \in Z : V(t,z) = 0\}$. Note that $V(t,z) \le 0$ for any $t \ge 0, z \in Z$.

The approach given above works only in the KPP case. To consider other types of nonlinear terms, let us introduce a new assumption. First, let us recall what is the physical sense of the coefficient $c^\epsilon(z,u)$ in (10.1). Of ourse, this depends on the physical (or biological, or chemical) problem under consideration. For example, in some biological models, $c^\epsilon(z,u)$ gives the rate of multiplication per unit for a given density of the "particles" u at a given point $z \in Z$. It characterizes the "convenience" of the environment and is called the fitness coefficient. If $c^\epsilon(z,u) > c_0 > 0$, then $u^\epsilon(t,z)$ is exponentially large as $\epsilon \downarrow 0$, provided that the initial condition is positive.

It is natural to assume in some problems that the fitness coefficient $c^\epsilon(x,u^\epsilon)$ depends on u^ϵ in a rough way: just large (exponentially) changes of the density influence the fitness coefficient. This can be expressed formally as follows:

$$c^\epsilon(z,u^\epsilon) = c(z, \epsilon \log u^\epsilon), \tag{10.2}$$

where $c(z,v)$ is a Lipschitz continuous bounded function. We assume, for brevity, that the initial function has an exponential form as well:

$$g(z) = g^\epsilon(z) = \exp\{\frac{\gamma(z)}{\epsilon}\},$$

where $\gamma(z)$ is a bounded uniformly continuous function. Then (10.1) becomes an equation for $v^\epsilon(t,z) = \epsilon \log u^\epsilon(t,x)$:

$$v^\epsilon(t,z) = \epsilon \log E_z^\epsilon \exp\{\frac{1}{\epsilon}[\gamma(Z_t^\epsilon) + \int_0^t c(Z_s^\epsilon, v^\epsilon(t-s, Z_s^\epsilon))ds]\}. \tag{10.3}$$

We are interested in the asymptotic behavior of $v^\epsilon(t,z)$ as $\epsilon \downarrow 0$. This is called large scale asymptotics for equation (10.1).

Consider the following equation.

$$v(t,z) = \sup_{\phi \in C_{0t}(Z), \phi_t = z} \{\gamma(\phi_0) + \int_0^t c(\phi_s, v(s, \phi_s))ds - S_{0t}(\phi)\}, \quad (10.4)$$

where $\frac{1}{\epsilon} S_{0t}(\phi)$ is the action functional for $(Z_t^\epsilon, P_z^\epsilon)$ as $\epsilon \downarrow 0$ in the uniform topology. Let us check that equation (10.4) has a unique continuous solution $v(t,z)$, $t \geq 0$, $z \in Z$. Set

$$M[v](t,z) = \sup_{\phi \in C_{0t}(Z), \phi_t = z} \{\gamma(\phi_0) + \int_0^t c(\phi_s, v(s, \phi_s))ds - S_{0t}(\phi)\}.$$

Denoting by K the Lipschitz constant for $c(z,v)$, we have:

$$| M[u](t,z) - M[v](t,z) | \leq | \sup_{\phi_t = z} [\gamma(\phi_0) + \int_0^t c(\phi_s, u(s, \phi_s))ds - S_{0t}(\phi)] -$$

$$- \sup_{\psi_t = z} [\gamma(\psi_0) + \int_0^t c(\psi_s, v(s, \psi_s))ds - S_{0t}(\psi)]$$

$$\leq \sup_{\phi_t = z} | \int_0^t [c(\phi_s, u(s, \phi_s))ds - c(\phi_s, v(s, \phi_s))]ds |$$

$$\leq K \sup_{\phi_t = z} \int_0^t | u(s, \phi_s) - v(s, \phi_s) | \, ds.$$

This inequality implies that

$$\max_{0 \leq s \leq t} \sup_{z \in Z} | M[u](z,s) - M[v](z,s) | \leq K \int_0^t (\max_{0 \leq s_1 \leq s} \sup_{z \in Z} | u(s_1, z) - v(s_1, z) |)ds.$$

Thus, for a sufficiently large power n, the operator M^n is a contraction in the space of continuous functions $u(s,z)$, $0 \leq s \leq T < \infty$, $z \in Z$, provided with the uniform metric. Therefore there exists a unique fixed point for M. This fixed point is the unique continuous solution of (10.4).

Let us now prove that $\lim_{\epsilon \downarrow 0} v^\epsilon(t,z) = v(t,z)$. Here $v^\epsilon(t,z)$ is the solution of (10.3) and $v(t,z)$ is the solution of (10.4). Subtracting (10.4) from (10.3), we have

$$v^\epsilon(t,z) - v(t,z) = [\epsilon \log E_z^\epsilon \exp\{\frac{1}{\epsilon}[\gamma(Z_t^\epsilon) + \int_0^t c(Z_s^\epsilon, v^\epsilon(t-s, Z_s^\epsilon))ds]\}]$$

$$- \epsilon \log E_z^\epsilon \exp\{\frac{1}{\epsilon}[\gamma(Z_t^\epsilon) + \int_0^t c(Z_s^\epsilon, v(t-s, z_s^\epsilon))ds]\}] +$$

$$+ [\epsilon \log E_z^\epsilon \exp\{\frac{1}{\epsilon}[\gamma(Z_t^\epsilon) + \int_0^z c(Z_s^\epsilon, v(t-s, Z_s^\epsilon))ds] - v(t,z)]$$

$$(10.5)$$

$$= \epsilon \log \frac{E_z^\epsilon \exp\{\frac{1}{\epsilon}[\gamma(Z_t^\epsilon) + \int_0^t c(Z_s^\epsilon, v^\epsilon(t-s, Z_s^\epsilon))ds]\}}{E_z^\epsilon \exp\{\frac{1}{\epsilon}[\gamma(Z_t^\epsilon) + \int_0^t c(Z_s^\epsilon, v(t-s, Z_s^\epsilon))ds]\}} + o_\epsilon(1).$$

Here we used the equality

$$
\begin{aligned}
v(t, z) &= \lim_{\epsilon \downarrow 0} \epsilon \log E_z^\epsilon \exp\{\frac{1}{\epsilon}[\gamma(Z_t^\epsilon) + \int_0^t c(Z_s^\epsilon, v(t - s, Z_s^\epsilon))ds\} \\
&= \sup_{\phi_0 = z} \{\gamma(\phi_t) + \int_0^t c(\phi_s, v(t - s, \phi_s))ds - S_{0t}(\phi)\}
\end{aligned}
$$

which follows from (10.4) and the properties of the action functional.

Consider now the first term of the right-hand side of (10.5). Denote by η_ϵ the random variable under the expectation sign in the denominator of this term. Then we can rewrite it as

$$
\epsilon \log \frac{E_z^\epsilon \eta_\epsilon \exp\{\frac{1}{\epsilon} \int_0^t [c(Z_s^\epsilon, v^\epsilon(t - s, Z_s^\epsilon)) - c(Z_s^\epsilon, v(t - s, Z_s^\epsilon))]ds\}}{E_z^\epsilon \eta_\epsilon}. \tag{10.6}
$$

Applying the Hölder inequality to (10.6), we derive from (10.5)

$$
\begin{aligned}
| v^\epsilon(t, z) - v(t, z) | &\leq o_\epsilon(1) + \epsilon \log \frac{(E_z^\epsilon \eta_\epsilon^p)^{1/p}}{E_z^\epsilon \eta_\epsilon} + \\
&\quad + \frac{\epsilon}{q} \log E_z^\epsilon \exp\{\frac{q}{\epsilon} \int_0^t [c(Z_s^\epsilon, v^\epsilon(t - s, Z_s^\epsilon)) - \\
&\quad - c(Z_s^\epsilon, v(t - s, Z_s^\epsilon))]ds\},
\end{aligned} \tag{10.7}
$$

where $p, q > 0$, $p + q = 1$. Let

$$
\mu_\epsilon(t) = \sup_{0 \leq s \leq t, z \in Z} | v^\epsilon(s, z) - v(s, z) |,
$$

$$
\mathcal{A}(p, \epsilon) = \epsilon \log \frac{(E_z^\epsilon \eta_\epsilon^p)^{1/p}}{E_z^\epsilon \eta_\epsilon}.
$$

The bound (10.7) implies:

$$
\mu_\epsilon(t) \leq [o_\epsilon(1) + \mathcal{A}(p, \epsilon)] + K \int_0^t \mu_\epsilon(s)ds,
$$

where K is the Lipschitz constant. Thus for any $t \geq 0$,

$$
\mu_\epsilon(t) \leq (o_\epsilon(1) + \mathcal{A}(p, \epsilon)) \exp\{Kt\}. \tag{10.8}
$$

Taking into account the fact that $\lim_{p \uparrow 1} \mathcal{A}(p, \epsilon) = 0$, for any $\epsilon > 0$, we conclude from (10.8) that $\lim_{\epsilon \downarrow 0} \mu_\epsilon(t) = 0$. Thus, we have the following result:

Theorem 10.1 *There exists a unique continuous solution $u^\epsilon(t, z)$ of equation (10.1) for any $\epsilon > 0$, and*

$$
\lim_{\epsilon \downarrow 0} \epsilon \log u^\epsilon(t, z) = \lim_{\epsilon \downarrow 0} v^\epsilon(t, z) = v(t, z)
$$

uniformly in $0 \leq t \leq T < \infty$, $z \in Z$, where $v(t, z)$ is the unique solution of (10.4).

Let, for example, $(Z_t^\epsilon, P_z^\epsilon)$ be the non-degenerate diffusion process in R^r governed by the operator

$$\epsilon L = \frac{\epsilon}{2} \sum_{i,j=1}^{r} a^{ij}(z) \frac{\partial^2}{\partial z^i \partial z^j}.$$

The normalized action functional $S_{0T}(\phi)$ for the family Z_t^ϵ, $0 \le t \le T$, in $C_{0T}(R^r)$ is given as

$$S_{0T}(\phi) = \begin{cases} \frac{1}{2} \int_0^T \sum_{i,j=1}^r a_{ij}(\phi) \dot\phi_s^i \dot\phi_s^j ds, & \phi_s \text{ is absolutely continuous;} \\ +\infty, & \text{for the rest of } C_{0T}(R^r), \end{cases}$$

$$(a_{ij}(z)) = (a^{ij}(z))^{-1}.$$

In this case the solution of equation (10.1) solves the Cauchy problem

$$\frac{\partial u^\epsilon(t,z)}{\partial t} = \epsilon L u^\epsilon(t,z) + \frac{1}{\epsilon} c(z, \epsilon \log u^\epsilon(t,z)) u^\epsilon(t,z), t > 0,\ z \in R^r, u^\epsilon(0,z)$$

$$= \exp\{\frac{1}{\epsilon}\gamma(z)\}.$$

The function $v^\epsilon(t,z) = \epsilon \log u^\epsilon(t,z)$ satisfies the equation

$$\frac{\partial v^\epsilon(t,z)}{\partial t} = \epsilon L v^\epsilon(t,z) + \frac{1}{2} \sum_{i,j=1}^{r} a^{ij}(z) \frac{\partial v^\epsilon}{\partial z^i} \frac{\partial v^\epsilon}{\partial z^j} + c(z, v^\epsilon), \qquad (10.9)$$

$$v^\epsilon(0,z) = \gamma(z).$$

One can expect that $v^\epsilon(t,z)$ converges as $\epsilon \downarrow 0$ to a solution of the degenerate problem:

$$\frac{\partial v(t,z)}{\partial t} = \frac{1}{2} \sum_{i,j=1}^{r} a^{ij}(z) \frac{\partial v}{\partial z^i} \frac{\partial v}{\partial z^j} + c(z, v), \qquad (10.10)$$

$$v(0,z) = \gamma(z).$$

The solution of equation (10.4) is a generalized solution of (10.10); (10.10) is the Hamilton-Jacobi equation for variational problem (10.4).

Consider the case $Z = R^1$, $a_{11}(x) = 1$, $c(x,v) = c(v)$. Then (10.4) has the form

$$v(t,x) = \sup_{\phi \in C_{0t}, \phi_t = x} \{\gamma(\phi_0) + \int_0^t c(v(s, \phi_s))ds - \frac{1}{2} \int_0^t |\dot\phi_s|^2 ds\}. (10.11)$$

If $\gamma(x) = \gamma = const$, then the solution of (10.11) is independent of x and satisfies the equation

$$v'(t) = c(v(t)),\ v(0) = \gamma. \qquad (10.12)$$

This equation plays the part of the local dynamics. The function $c(v)$ shown in Fig. 26a corresponds to the KPP case: equation (10.12) has one stable equilibrium point at $v = v_{cr}$ and an unstable rest point at $-\infty$. Note that $v = -\infty$ corresponds to $u = 0$.

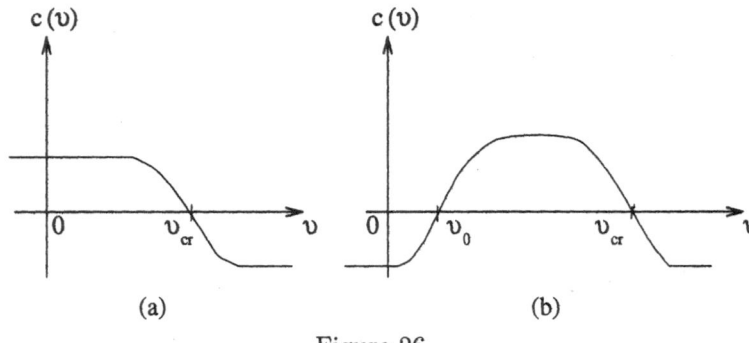

Figure 26

The function $c(v)$ shown in Fig. 26b corresponds to the bistable case: stable equilibriums at v_{cr} and at $-\infty$ separated by the unstable rest point at $v = v_0$. Equation (10.11) defines a nonlinear semigroup $T_t[\gamma] = v(t, x)$. One can consider stationary points of this semigroup (solutions of (10.11) which are independent of t) as well as running wave type solutions.

Equation (10.11) can be considered, actually, for any initial function $\gamma(x)$ taking values in $[-\infty, A]$ for some finite A. One can introduce "the fundamental" solution $v_y(t, x)$ of (10.11): $v_y(t, x)$ is the solution of (10.11) with $\gamma(x) = \gamma_y(x) = 0$ for $x = y$ and $\gamma_y(x) = -\infty$ for $x \neq y$. Then

$$v_y(t, x) = \sup\{\int_0^t [c(v_y(s, \phi_s)) - \frac{\dot{\phi}_s^2}{2}]ds : \phi_0 = y, \ \phi_t = x\}.$$

The Hamilton-Jacobi equation for $v_y(t, x)$ holds:

$$\frac{\partial v_y(t, x)}{\partial t} = \frac{1}{2}(\frac{\partial v_y}{\partial x})^2 + c(v_y),$$

$\lim_{t \downarrow 0} v_y(t, x) = 0$ for $x = y$, and $\lim_{t \downarrow 0} v_y(t, x) = -\infty$ for $x \neq y$.

We mention one more property of equation (10.11): if $v_1(t, x)$ and $v_2(t, x)$ are solutions of (10.11) with initial functions $\gamma_1(x)$ and $\gamma_2(x)$ respectively, then the solution of (10.11) with $\gamma(x) = \gamma_1(x) \vee \gamma_2(x)$ is $v(t, x) = v_1(t, x) \vee v_2(t, x)$.

Consider now another example where the large scale approximation can be used. Let $Z_t^\epsilon = (X_t^\epsilon, Y_t^\epsilon)$ be the Markov process in the strip $\sqcap = \{(x, y) \in R^2 : -\infty < x < \infty, |y| \leq 1\}$ governed by the operator

$$L^\epsilon = \frac{1}{2\epsilon}\frac{\partial^2}{\partial y^2} + b(y)\frac{\partial}{\partial x}$$

inside $\sqcap$ and having normal reflection on the boundary $\partial \sqcap$. This means that $Y_t^\epsilon = \bar{Y}_{t/\epsilon}$, where $\bar{Y}_t$ is the Wiener process in $[-1, 1]$ with reflection in the ends of the interval, and

$$X_t^\epsilon \;=\; x + \int_0^t b(Y_s^\epsilon)ds. \tag{10.13}$$

We assume that $b(y), y \in [-1, 1]$ is a bounded, continuously differentiable function. Let the evolution of the particle density $u^\epsilon(t, x, y)$, $(x, y) \in \sqcap$, $t \geq 0$, be described by the equation

$$u^\epsilon(t, x, y) = E_{x,y} \exp\{\frac{1}{\epsilon}[\gamma(X_t^\epsilon) + \int_0^t c(X_s^\epsilon, \epsilon \log u^\epsilon(t - s, X_s^\epsilon, Y^\epsilon, s))ds]\}. \tag{10.14}$$

Assume, for brevity, that $\gamma(x)$, $x \in R^1$, has a bounded derivative. Equation (10.14) can be written in the differential form

$$\frac{\partial u^\epsilon(t, x, y)}{\partial t} \;=\; L^\epsilon u^\epsilon + \frac{1}{\epsilon}c(x, \epsilon \log u^\epsilon)u^\epsilon,$$

$$t > 0, \; x \in R^1, |\, y\,| < 1,$$

$$\frac{\partial u^\epsilon(t, x, \pm 1)}{\partial y} \;=\; 0, \; u^\epsilon(0, x, y) = \exp\{\frac{\gamma(x)}{\epsilon}\}. \tag{10.15}$$

The particle transport here is described, at least asymptotically, by the component X_t^ϵ of the Markov process $(X_t^\epsilon, Y_t^\epsilon)$. Therefore, as we will see, the logarithmic asymptotic of $u^\epsilon(t, x, y)$ as $\epsilon \downarrow 0$ is defined by the action functional for the process X_t^ϵ.

Let $v^\epsilon(t, x, y) = \epsilon \log u^\epsilon(t, x, y)$. Then $v^\epsilon(t, x, y)$ is the solution of the problem

$$\frac{\partial v^\epsilon(t, x, y)}{\partial t} \;=\; \frac{1}{2\epsilon}\frac{\partial^2 v^\epsilon}{\partial y} + \frac{1}{2\epsilon^2}\left(\frac{\partial v^\epsilon}{\partial y}\right)^2 + b(y)\frac{\partial v^\epsilon}{\partial x} + c(x, v^\epsilon),$$

$$t > 0, \; x \in R^1, |\, y\,| < 1,$$

$$\tag{10.16}$$

$$\frac{\partial v^\epsilon(t, x, \pm 1)}{\partial y} \;=\; 0, \; v^\epsilon(0, x, y) = \gamma(x)$$

To calculate $\lim_{\epsilon \downarrow 0} v^\epsilon(t, x, y)$, we need the action functional for the family X_t^ϵ, $0 \leq t \leq T$, as $\epsilon \downarrow 0$. It can be calculated as follows (See [F2], [FW1]): consider the eigenvalue problem

$$\frac{1}{2}\phi''(y) + \alpha b(y)\phi(y) \;=\; \lambda\phi(y), \; |\, y\,| < 1, \tag{10.17}$$

$$\phi'(\pm 1) \;=\; 0.$$

Here $\alpha \in R^1$ is a parameter. Let $\lambda = \lambda(\alpha)$ be the eigenvalue corresponding to the positive eigenfunction of problem (10.17). Such a $\lambda(\alpha)$ exists and is unique; moreover, it is a differentiable and convex function. Denote by $L(\beta)$ the Legendre transformation of $\lambda(\alpha)$:

$$L(\beta) = \sup_{\alpha \in r^1} [\alpha\beta - \lambda(\alpha)], \ \beta \in R^1.$$

Then the action functional for the family X_t^ϵ, $0 \le t \le T$, in $C_{0T}(R^1)$ is equal to $\epsilon^{-1} S_{0T}(\phi)$,

$$S_{0T}(\varphi) = \begin{cases} \int_0^T L(\dot{\phi}_s)ds, & \text{for absolutely continuous } \phi, \ \phi_0 = X_0^\epsilon = x; \\ +\infty, & \text{for the rest of } C_{0T}(R^r). \end{cases}$$

Consider the equation:

$$v(t,x) = \sup_{\phi \in C_{0t}(R^1), \phi_t = x} \{\gamma(\phi_0) + \int_0^t [c(\phi_s, v(s, \phi_s)) - L(\dot{\phi}_s)]ds \qquad (10.18)$$

Since we assume that $c(x,v)$ is Lipschitz continuous, equation (10.18) has a unique continuous solution. It is not difficult to check that $\max_{y_1, y_2 \in [-1,1]} |\ v^\epsilon(t, x, y_1) - v^\epsilon(t, x, y_2)\ | \to 0$ as $\epsilon \downarrow 0$. Then one can prove, in the same way as Theorem 10.1 was proved, that

$$\lim_{\epsilon \downarrow 0} v^\epsilon(t, x, y) = v(t, x, y).$$

Note that now it is impossible to obtain the Hamilton-Jacobi equation for $v(t,x)$ just setting ϵ to 0 in equation (10.16) for $v^\epsilon(t,x)$, as can be done in the case of equations (10.9) and (10.10).

Let us demonstrate now how the large scale approximation works in the case of a non-local nonlinear term. In some models, it is natural to assume that the "fitness coefficient" c^ϵ depends not just on the density at one point but on the whole amount of particles:

$$\frac{\partial u^\epsilon(t,x)}{\partial t} = \frac{\epsilon}{2}\Delta u^\epsilon(t, x) + \frac{1}{\epsilon}c^\epsilon\left(x, \int_{R^r} u^\epsilon(t, y)dy\right)u^\epsilon(t, x), \ t > 0, \ x \in R^r. \quad (10.19)$$

The large scale approximation now means that

$$c^\epsilon\left(x, \int_{R^r} u^\epsilon(t, y)dy\right) = c\left(x, \epsilon\log\int_{R^r} u^\epsilon(t, y)dy\right),$$

where $c(x,v)$ is a bounded Lipschitz continuous function. One can check that problem (10.19) has a unique solution for any bounded continuous initial function $u^\epsilon(0, x) = g^\epsilon(x) = \exp\{\frac{\gamma(x)}{\epsilon}\}$; $\epsilon > 0$ is a parameter.

Let $v^\epsilon(t,x) = \epsilon \log u^\epsilon(t,x)$ and consider the equation:

$$v(t,x) = \sup_{\phi \in C_{0t}(R^r), \phi_t = x} \{\gamma(\phi_0) + \int_0^t [c(\phi_s, \sup_{y \in R^r} v(s,y)) - \frac{|\dot\phi_s|^2}{2}]ds\}. \qquad (10.20)$$

Again, since $c(x,v)$ is Lipschitz continuous, (10.20) has a unique continuous solution. Using the large deviation principle, one can prove, similarly to Theorem 10.1, that

$$\lim_{\epsilon \downarrow 0} v^\epsilon(t,x) = v(t,x).$$

Equation (10.20) is actually simpler than (10.4). Let $c(x,v) = c^{(1)}(x) + c^{(2)}(v)$. Then (10.20) can be solved "almost explicitly": put $m(t) = \sup_{x \in R^r} v(t,x)$. Then (10.20) implies

$$\frac{dm(t)}{dt} = \dot m(t) = c^{(2)}(m(t)) + A'(t), \quad m(0) = \bar\gamma = \sup_{x \in R^r} \gamma(x),$$

where

$$A(t) = \sup_{x \in R^r} A(t,x), \quad A(t,x) = \sup_{\phi \in C_{0t}(R^r), \phi_t = x} \{\gamma(\phi_0) + \int_0^t [c^{(1)}(\phi_s) - \frac{|\dot\phi_s|^2}{2}]ds\}.$$

Note that $\lim_{t \to \infty} A'(t) = \sup_{x \in R^r} c^{(1)}(x) = \bar c^{(1)}$. Therefore $m(t)$, for large t, behaves as the solution of the equation

$$\dot{\bar m}(t) = c^{(2)}(\bar m(t)) + \bar c^{(1)}.$$

For $v(t,x)$, we have an expression:

$$v(t,x) = \int_0^t c^{(2)}(m(s))ds + A(t,x) = m(t) + A(t,x) - A(t) + \bar\gamma. \qquad (10.21)$$

Now we will briefly consider the large scale approximation for RDE-systems. First, examine the non-local system

$$\frac{\partial u_1^\epsilon(t,x)}{\partial t} = \frac{\epsilon D_1}{2}\Delta u_1^\epsilon(t,x) + \frac{1}{\epsilon}c_1(x,\epsilon\log\int_{R^r} u_1^\epsilon(t,y)dy, \epsilon\log\int_{R^r} u_2^\epsilon(t,y)dy)u_1^\epsilon(t,x)$$

$$\frac{\partial u_2^\epsilon(t,x)}{\partial t} = \frac{\epsilon D_2}{2}\Delta u_2^\epsilon(t,x) + \frac{1}{\epsilon}c_2(x,\epsilon\log\int_{R^r} u_1^\epsilon(t,y)dy, \epsilon\log\int_{R^r} u_2^\epsilon(t,y)dy)u_2^\epsilon(t,x)$$

$$u_1^\epsilon(0,x) = \exp\{\frac{\gamma_1(x)}{\epsilon}\} \quad u_2^\epsilon(0,x) = \exp\{\frac{\gamma_2(x)}{\epsilon}\}.$$

One can prove that $v_i^\epsilon(t,x) = \epsilon \log u_i^\epsilon(t,x) \to v_i(t,x)$ as $\epsilon \downarrow 0$, $i = 1, 2$, and the functions $v_1(t,x), v_2(t,x)$ satisfy the equations

$$v_i(t,x) \;=\; \sup_{\phi \in C_{0t}(R^r),\phi_t = x} \{ \gamma_i(\phi_0) +$$

$$+ \int_0^t [c_i(\phi_s, \sup_{y \in R^r} v_1(s,y), \sup_{y \in R^r} v_2(s,y) - \frac{|\dot{\phi}_s|^2}{2D_i}]ds\}, \quad (10.22)$$

where $t \geq 0$, $x \in R^r$, $i = 1, 2$.

Assume that $c_i(x,v) = c_i^{(1)}(x) + c_i^{(2)}(v)$, $i = 1, 2$, and let $m_i(t) = \sup_{x \in R^r} v_i(t,x)$. Then, similar to the case of one equation

$$\dot{m}_i(t) = c_i^{(2)}(m_1(t), m_2(t)) + \mathcal{A}_i'(t), \quad m_i(0) = \bar{\gamma}_i = \sup_{x \in R^r} \gamma_i(x),$$

$$\mathcal{A}_i(t) = \sup_{x \in R^r} \mathcal{A}_i(t,x), \quad \mathcal{A}_i(t,x) = \sup_{\phi \in C_{0t}(R^r),\phi_t = x} \{\gamma_i(\phi_0) + \int_0^t [c_i^{(1)}(\phi_s) - \frac{|\dot{\phi}_s|^2}{2D_i}]ds\},$$

$t > 0$, $i = 1, 2$. Again, the behavior of $(m_1(t), m_2(t))$ for large t can be described by the equations

$$\dot{m}_i(t) = c_i^{(2)}(m_1(t), m_2(t)) + \bar{c}_i^{(1)},$$

where $\bar{c}_i^{(1)} = \sup_{x \in R^r} c_i^{(1)}(x)$, $i = 1, 2$. One can write expressions similar to (10.21) for $v_i(t,x)$.

Remark. If $c^\epsilon(\cdot, \cdot)$ in (10.19) is independent of ϵ, the solution of (10.19) with initial conditions $u^\epsilon(0,x) = g(x) \geq 0$, under some additional assumptions, behaves as a running impulse as $\epsilon \downarrow 0$. The motion of that impulse can be described using the large deviation principle (See [F9]).

Finally, I will mention how the large scale approximation works for a RDE-system with a strong binding between the equations:

$$\frac{\partial u_k^\epsilon(t,x)}{\partial t} \;=\; \frac{\epsilon D_k}{2} \Delta u_k^\epsilon + \frac{1}{\epsilon} \sum_{j=1}^n c_{kj}(x)(u_j^\epsilon - u_k^\epsilon) +$$

$$+ \frac{1}{\epsilon} c_k(x; \epsilon \log u_1^\epsilon(t,x), \ldots, \epsilon \log u_n^\epsilon(t,x))u_k^\epsilon, \quad (10.23)$$

$$t > 0, \; x \in R^r$$

$$u_k^\epsilon(0,x) \;=\; \exp\{\frac{\gamma_k(x)}{\epsilon}\}, \; k = 1, \ldots, n.$$

Let $(X_t^\epsilon, \nu_t^\epsilon)$ be the process in $R^r \times \{1, \ldots, n\}$ corresponding to the linear part of the system (10.23), $v_k^\epsilon(t,x) = \epsilon \log u_k^\epsilon(t,x)$. We assume that the coefficients $c_{kj}(x)$ in (10.23) are strictly positive. Therefore ν_t^ϵ is a rapidly changing process as $\epsilon \downarrow 0$.

Because of this, the logarithmic asymptotic of all functions $u_k^\epsilon(t,x)$ as $\epsilon \downarrow 0$ turns out the same:

$$\lim_{\epsilon \downarrow 0} v_k^\epsilon(t,x) = v'(t,x), \ k = 1,\ldots,n.$$

To describe the limiting function $v(t,x)$, we need the action functional $\frac{1}{\epsilon}S_{0t}(\phi,\mu)$ for the family $(X_t^\epsilon, Z_t^\epsilon)$, where $Z_t^\epsilon = (Z_t^{\epsilon,1},\ldots,Z_t^{\epsilon,n})$ and $Z_t^{\epsilon,k}$ is the occupation time at the state k for ν_s^ϵ:

$$Z_t^{\epsilon,k} = \int_0^t \chi_k(\nu_s^\epsilon)ds,$$

χ_k is the indicator function of the state $k \in \{1,\ldots,n\}$. It was shown in [FL1], that

$$S_{0t}(\varphi) = \begin{cases} \frac{1}{2}\int_0^t \eta(\phi_s,\dot\phi_s,\dot\mu_s)ds, \ \phi \in C_{0t}(\mathbf{R}^r), \phi_0 = X_0^\epsilon = x, \\ \phi \text{ is absolutely cont. and } \mu \in M_{0t}; \\ \\ +\infty, \text{otherwise.} \end{cases}$$

M_{0t} is the set of functions $\mu(s) = (\mu_1(s),\ldots,\mu_n(s)), 0 \le s \le t$, such that $\mu_i(0) = 0, \mu_i(s)$ are non-decreasing functions and $\sum_{i=1}^n \mu_i(s) = s$.

To define the function $\eta(x,q,\beta), \ x \in R^r, \ q \in R^r, \ \beta \in R^n$, consider the matrix $(A_{km}(x,p,\alpha))$,

$$\begin{aligned} A_{km} &= c_{km}(x), \ k \neq m, \\ A_{kk} &= -\sum c_{kj}(x) + \frac{D_k}{2}|p|^2 + \alpha_k, \end{aligned}$$

$x,p \in R^r, \ \alpha \in R^n$. Let $\lambda = \lambda(x,p,\alpha)$ be the first eigenvalue of the matrix $(A_{km}(x,p,\alpha))$. Then $\eta(x,q,\beta)$ is the Legendre transformation of $\lambda(x,p,\alpha)$:

$$\eta(x,q,\beta) = \sup_{\alpha \in R^n, p \in R^r} (\alpha \cdot \beta + p \cdot q - \lambda(x,p,\alpha)).$$

Now, we can write down the equation for $v(t,x)$:

$$\begin{aligned} v(t,x) = &\sup_{\phi \in C_{0t}(R^n),\phi_t=x} [\bar\gamma(\phi_0) + \\ &\int_0^t [\sum_{k=1}^n \hat c_k(\phi_s,v(s,\phi_s))\dot\mu_k(s) - \eta(\phi_s,\dot\phi_s,\dot\mu_s)]ds], \quad (10.24) \end{aligned}$$

where $\hat c_k(x,v) = c_k(x,v,\ldots,v), \bar\gamma(x) = \max_{1 \le k \le n}\gamma_k(x)$. Equation (10.24) has a unique solution under our assumptions.

11 Homogenization in PDE's and in Stochastic Processes

Consider a second order elliptic operator with 1-periodic coefficients

$$L = \frac{1}{2} \sum_{i,j=1}^{r} a^{ij}(x) \frac{\partial^2}{\partial x^i \partial x^j} + \sum b^i(x) \frac{\partial}{\partial x^i}, \quad x \in \mathbf{R}^r.$$

Denote by L^ϵ the differential operator obtained from L when we replace x by $\frac{x}{\epsilon}$ in the coefficients, $\epsilon > 0$. The operator L^ϵ has fast oscillating coefficients for $\epsilon \ll 1$. Let X_t^ϵ be the diffusion process in $\mathbf{R}^r$ governed by L^ϵ. One can expect that for small $\epsilon > 0$ this process is close, in sense of weak convergence topology, for example, to a diffusion process governed by an operator $\bar{L}$ with constant coefficients In terms of PDE's this means that the solutions of initial-boundary problems for the equation $\frac{\partial u^\epsilon(t,x)}{\partial t} = L^\epsilon u^\epsilon$ or solutions of boundary problems for the stationary equation $L^\epsilon u^\epsilon(x) = f(x)$, converge as $\epsilon \downarrow 0$ to the solutions of the corresponding problems for the equation with constant coefficients. The problem of finding such an operator $\bar{L}$ and proving the convergence is a typical homogenization problem. It was considered in [F1]. Many interesting results concerning the homogenization appeared since then. Equations in self-adjoint form with rapidly oscillating periodic coefficients were studied in [BLP]. The case of space homogeneous random coefficients was considered [K], [PV]. Media with periodically and randomly distributed "holes" with various boundary conditions were considered as well.

Let us start with the problem formulated above. The operator L can be considered in the space of functions on the 1-torus T. Denote by $\hat{X}_t$ the diffusion process on T, corresponding to L. Define $\hat{\hat{X}}_t$ as the process on T governed by the 'shortened' operator $\frac{1}{2} \sum_{i,j=1}^{r} a^{ij}(x) \frac{\partial^2}{\partial x^i \partial x^j}$, which is the main part of L. Let $m(x)$ be the density of the normalized invariant measure of the process $\hat{\hat{X}}_t$ on the torus T. Note that $m(x)$ is the unique solution of the problem

$$\sum_{i,j=1}^{r} \frac{\partial^2}{\partial x^i \partial x^j} \left(a^{ij}(x) m(x) \right) = 0, \quad x \in T, \quad \int_T m(x)\, dx = 1.$$

Define

$$\bar{a}^{ij} = \int_T a^{ij}(x) m(x)\, dx, \quad \bar{b}^i(x) = \int_T b^i(x) m(x)\, dx.$$

Let $\bar{X}_t$ be the Markov Gaussian process in $\mathbf{R}^r$ governed by the operator

$$\bar{L} = \frac{1}{2} \sum_{i,j=1}^{r} \bar{a}^{ij} \frac{\partial^2}{\partial x^i \partial x^j} + \sum_{i=1}^{r} \bar{b}^i \frac{\partial}{\partial x^i}.$$

We will prove now that the processes X_t^ϵ, $0 \le t \le T$, converge weakly in $C_{0T}(\mathbf{R}^r)$ to the process $\bar{X}_t$ as $\epsilon \downarrow 0$.

First, note that the family $\{X_t^\epsilon\}$ is tight in the topology of weak convergence in $C_{0T}(\mathbf{R}^r)$: this follows from the boundedness of the coefficients which is uniform in $\epsilon \in (0,1]$. Thus, to prove the weak convergence, one only needs to prove the convergence of finite-dimensional distributions.

Let $\tilde{X}_t^\epsilon$ be the process in $\mathbf{R}^r$ governed by the operator

$$\tilde{L}^\epsilon = \frac{1}{2} \sum_{i,j=1}^{r} a^{ij}(x) \frac{\partial^2}{\partial x^i \, \partial x^j} + \epsilon \sum_{i=1}^{r} b^i(x) \frac{\partial}{\partial x^i}.$$

One can easily check that $\epsilon^{-1} X_t^\epsilon$, $X_0^\epsilon = x$, has the same distribution in $C_{0T}(\mathbf{R}^r)$ as $\tilde{X}_{t/\epsilon^2}$ $\tilde{X}_0^\epsilon = \epsilon^{-1} x$. The operator $\tilde{L}^\epsilon$ can be considered on the torus T. Denote by $\tilde{\tilde{X}}_t^\epsilon$ the process on T corresponding to $\tilde{L}^\epsilon$, and let $m^\epsilon(x)$ be its normalized invariant density. It is easy to prove that $m^\epsilon(x) \to m(x)$ uniformly in $x \in \mathbf{R}^r$ as $\epsilon \downarrow 0$, and that for any bounded measurable function $f(x)$, $x \in T$,

$$\left| E_x f\left(\tilde{\tilde{X}}_{t/\epsilon^2}^\epsilon\right) - \int_T f(x) m(x) \, dx \right| \to 0 \tag{11.1}$$

as $\epsilon \downarrow 0$ uniformly in $x \in T$ and $t \ge t_0 > 0$. Processes $f\left(\tilde{X}_t^\epsilon\right)$ and $f\left(\tilde{\tilde{X}}_t^\epsilon\right)$, $0 \le t \le T$, for any 1-periodic function $f(x)$, $x \in \mathbf{R}^r$, have the same distribution so that we derive from (11.1), that for any 1-periodic $f(x), x \in \mathbf{R}^r$,

$$\lim_{\epsilon \downarrow 0} E_x f\left(\tilde{X}_{t/\epsilon^2}^\epsilon\right) = \int_T f(x) m(x) \, dx \tag{11.2}$$

uniformly in $x \in \mathbf{R}^r$, $t \ge t_0 > 0$.

Let us for brevity consider the case $r = 1$ and one-dimensional distributions. The general case is similar.

Using the self-similarity of the Wiener process, we have

$$\begin{aligned}
X_t^\epsilon &= x + \int_0^t \sigma\left(\epsilon^{-1} X_s^\epsilon\right) dW_s + \int_0^t b\left(\epsilon^{-1} x_s^\epsilon\right) ds \\
&\sim \tilde{W}\left(\int_0^t a\left(\epsilon^{-1} X_s^\epsilon\right) ds\right) + \int_0^t b\left(\epsilon^{-1} X_s^\epsilon\right) ds + x \qquad (11.3) \\
&\sim \tilde{W}\left(\int_0^t a\left(\tilde{X}_{s/\epsilon^2}^\epsilon\right) ds\right) + \int_0^t b\left(\tilde{X}_{s/\epsilon^2}^\epsilon\right) ds + x.
\end{aligned}$$

Here $\sigma(x) = \sqrt{a(x)}$; the sign '$\sim$' means equivalence of the distributions; W_s and $\tilde{W}_s$ are one-dimensional Wiener processes. Since $a(x)$ and $b(x)$ are 1-periodic functions, we can replace in the right hand side of (11.3) the processes $\tilde{X}_{s/\epsilon^2}^\epsilon$ by $\tilde{\tilde{X}}_{s/\epsilon^2}^\epsilon$.

Then, using the ergodic theorem for the process $\tilde{X}_t^\epsilon$ on the torus and (11.2), we derive that the right side of (11.3) converges in distribution as $\epsilon \downarrow 0$ to

$$x + \tilde{W}_{t\bar{a}} + t\bar{b} = \bar{X}_t. \qquad QED.$$

Let us consider, for example, one dimensional process without drift:

$$X_t^\epsilon - x = \int_0^t \sigma\left(\epsilon^{-1}X_s^\epsilon\right) dW_s, \quad \sigma(x) = \sigma(x+1) > 0.$$

Then the invariant density is equal to

$$m(x) = \frac{1}{\sigma^2(x)} \left(\int_0^1 \frac{dx}{\sigma^2(x)}\right)^{-1},$$

and we conclude that the "effective" diffusion is equal to

$$\bar{a} = \int_0^1 \sigma^2(x)m(x)\, dx = \left(\int_0^1 \frac{dx}{\sigma^2(x)}\right)^{-1}.$$

In a similar way, one can consider the processes governed by the elliptic operators in the divergent form with fast oscillating periodic coefficients. We restrict ourselves to the one-dimensional case. Let X_t^ϵ correspond to the operator $L: Lv(x) = \frac{1}{2}\frac{d}{dx}(a(x/\epsilon)\frac{dv(x)}{dx})$. Again, we assume that $a(x)$ is positive, 1-periodic and smooth enough. The trajectories X_t^ϵ, starting at $x \in R^1$, satisfy the equation

$$X_t^\epsilon = x + \int_0^t \sqrt{a(\epsilon^{-1}X_s^\epsilon)}dW_s + \frac{1}{2\epsilon}\int_0^t a'(\epsilon^{-1}X_s^\epsilon)ds.$$

At first glance, the last term appears to have order ϵ^{-1}. However, this is not correct. Let

$$u(x) = \int_0^x \frac{(\overline{a^{-1}})^{-1} - a(y)}{a(y)}dy, \quad \overline{a^{-1}} = \int_0^1 \frac{dy}{a(y)}.$$

The function $u(x)$ satisfies the equation $L^1u(x) = \frac{1}{2}(a(x)u'(x))' = -\frac{1}{2}a'(x)$ and the condition $\int_0^1 u(x)dx = 0$. Applying the Itô formula, we have

$$\epsilon u(\epsilon^{-1}X_t^\epsilon) - \epsilon u(\epsilon^{-1}x) = \int_0^t u'(\epsilon^{-1}X_s^\epsilon)\sqrt{a(\epsilon^{-1}X_s^\epsilon)}dW_s - \frac{1}{2\epsilon}\int_0^t a'(\epsilon^{-1}X_s^\epsilon)ds.$$

Combining this equality and the equation for X_t^ϵ, we derive

$$X_t^\epsilon - x = \int_0^t [\sqrt{a(\epsilon^{-1}X_s^\epsilon)} + u'(\epsilon^{-1}X_s^\epsilon)\sqrt{a(\epsilon^{-1}X_s^\epsilon)}]dW_s - \epsilon[u(\epsilon^{-1}X_t^\epsilon) - u(\epsilon^{-1}x)].$$

The last term here tends to zero as $\epsilon \downarrow 0$ since $u(x), x \in R^1$, is a bounded function. Using the self-similarity of the Wiener process, we can replace the first term by

$$\tilde{W}\left(\int_0^t a(\epsilon^{-1}X_s^\epsilon)(1 + u'(\epsilon^{-1}X_s^\epsilon))^2 ds\right),$$

where $\tilde{W}_t$ is a Wiener process in R^1. Since the function $a(x)(1+u'(x))$ is 1-periodic,

$$\int_0^t a(\epsilon^{-1}X_s^\epsilon)(1 + u'(\epsilon^{-1}X_s^\epsilon))^2 ds \to t\int_0^1 a(x)(1 + u'(x))^2 dx = t\hat{a}.$$

Recalling the expression for $u(x)$, we derive that $\hat{a} = (\overline{a^{-1}})^{-1} = (\int_0^1 \frac{dx}{a(x)})^{-1}$. One can conclude from these bounds that the process X_t^ϵ converges weakly in $C_{0T}(R^1)$ to the process governed by the operator $\hat{L} = \frac{\hat{a}}{2}\frac{d^2}{dx^2}$.

An important question is: what are the most general conditions on the coefficients of an elliptic operator

$$L^\epsilon = \frac{1}{2}\sum_{i,j=1}^r a_\epsilon^{ij}(x)\frac{\partial^2}{\partial x^i \partial x^j} + \sum_{i=1}^r b_\epsilon^i(x)\frac{\partial}{\partial x^i}$$

providing weak convergence of corresponding processes.

It follows from the above mentioned results that if $a_\epsilon^{ij}(x) = a^{ij}\left(\frac{x}{\epsilon}\right)$, $b_\epsilon^i(x) = b^i\left(\frac{x}{\epsilon}\right)$ and $a^{ij}(x)$ and $b^i(x)$ are 1-periodic then the corresponding processes converge to the process governed by $\bar{L}$. It is not difficult to generalize this result to the case when $a_\epsilon^{ij}(x) = a^{ij}\left(x, \frac{x}{\epsilon}\right)$, $b_\epsilon^i(x) = b^i\left(x, \frac{x}{\epsilon}\right)$ where $a^{ij}(x,y), b^i(x,y)$ are smooth functions 1-periodic in y. Then the limiting process is governed by the operator

$$\bar{L} = \frac{1}{2}\sum \bar{a}^{ij}(x)\frac{\partial^2}{\partial x^i \partial x^j} + \sum \bar{b}(x)\frac{\partial}{\partial x^i},$$

$$\bar{a}^{ij}(x) = \int_T a^{ij}(x,y)m_x(y)dy, \quad \bar{b}^i(x) = \int_T b^i(x,y)m_x(y)\,dy,$$

where $m_x(y)$ is the invariant density of the process on the torus governed by

$$L_x = \frac{1}{2}\sum_{i,j=1}^r a^{ij}(x,y)\frac{\partial^2}{\partial y^i \partial y^j},$$

$x \in R^r$ is a parameter.

One can weaken the conditions on the drift term: if, say, the $a_\epsilon^{ij}(x) = a^{ij}(x)$ are independent of ϵ and if the $b_\epsilon^i(x)$ are bounded and converge weakly in the sense of generalized functions to $b^i(x)$ as $\epsilon \downarrow 0$, then the corresponding stochastic processes converge weakly in the space of continuous functions on any finite time interval to the process governed by the limiting operator.

A complete, in a sense, solution of the problem in the one-dimensional case has been found recently in [FW4].

It is well known that continuous one-dimensional Markov processes can be described by the operators $D_\vartheta D_u$ [Fe]. Here $u(x)$ and $\vartheta(x)$ are strictly increasing functions, $u(x)$ is continuous, $\vartheta(x)$ is right continuous; the differentiation of a monotone function is defined as follows:

$$D_u f(x) = \lim_{\Delta \to 0} \frac{f(x + \Delta) - f(x)}{u(x + \Delta) - u(x)}.$$

The functions $u(x)$ and $v(x)$ are defined up to an additive constant; the pairs (u, v) and $(Au, A^{-1}v)$, $A > 0$, correspond to the same process. It is easy to calculate that for the process governed by the operator

$$L = a(x)\frac{d^2}{dx^2} + b(x)\frac{d}{dx}$$

the corresponding $u(x)$ and $\vartheta(x)$ are

$$
\begin{aligned}
u(x) &= \int_0^x dy \cdot \exp\left\{-\int_0^y \frac{b(z)}{a(z)}\, dz\right\}, \\
\vartheta(x) &= \int_0^x \frac{dy}{a(y)} \exp\left\{\int_0^y \frac{b(z)}{a(z)}\, dz\right\}.
\end{aligned}
\tag{11.4}
$$

Then if we have a family of operators L^ϵ, one can consider the family $(u^\epsilon(x), \vartheta^\epsilon(x))$. It is proved in [FW4], that the following conditions are sufficient and necessary for the weak convergence of processes X_t^ϵ governed by the operator $D_{\vartheta^\epsilon} D_{u^\epsilon}$ to the process X_t corresponding to $D_\vartheta D_u$:

$u^\epsilon(x) \to u(x)$ at any x,
$\vartheta^\epsilon(x) \to \vartheta(x)$ at any continuity point of the limiting function $\vartheta(x)$

(after a proper choice of the functions u, v connected with the above mentioned non-uniqueness).

In particular, if we are interested in the convergence to a space homogeneous Markov process, then the limiting $u(x)$ and $\vartheta(x)$ should be of the form

$$u(x) = \frac{a}{b}\left(1 - e^{-\frac{bx}{a}}\right), \quad \vartheta(x) = \frac{1}{b}\left(e^{\frac{bx}{a}} - 1\right) \quad \text{if } b \neq 0,$$

$$u(x) = x, \quad \vartheta(x) = \frac{x}{a} \quad \text{if } b = 0.$$

Here a and b are correspondingly the diffusion and the drift coefficients of the limiting process.

All homogenization results concerning the weak convergence for one-dimensional processes follow from the result formulated above. Consider, for example, a diffusion process in a random medium. Let the coefficients of the operator L have the form

$$a^\epsilon(x) = a(x/\epsilon), \ b^\epsilon(x) = b(x/\epsilon) + \frac{1}{\epsilon}\tilde{b}(x/\epsilon).$$

The last term in the expression for $b^\epsilon(x)$ is included to incorporate the self-adjoint operators with rapidly oscillating coefficients. Let $(a(x), b(x), \tilde{b}(x))$, $x \in R^1$ be a space homogeneous ergodic random field. Assume that $a(x) > 0$ with probability 1, $E(\tilde{b}(x)/a(x)) = 0$, and that the spectral measure $F(d\lambda)$ of the field $\tilde{b}(x)/a(x)$ satisfies the condition

$$\int_{-\infty}^{\infty} \frac{F(d\lambda)}{\lambda^2} < \infty.$$

Then the space homogeneous random field $\zeta(x)$, $x \in R^1$, exists such that

$$\zeta'(x) = \frac{d\zeta(x)}{dx} = \frac{\tilde{b}(x)}{a(x)}.$$

Under these conditions, one can write down the following expressions for the corresponding functions $u^\epsilon(x), v^\epsilon(x)$:

$$u^\epsilon(x) = \int_0^x \exp\{-\int_0^y \frac{b(z/\epsilon)}{a(z/\epsilon)}dz - \zeta(y/\epsilon)\}dy,$$

$$u^\epsilon(x) = \int_0^x \exp\{\int_0^y \frac{b(z/\epsilon)}{a(z/\epsilon)}dz + \zeta(y/\epsilon)\}\frac{dy}{a(y/\epsilon)}.$$

It is easy to calculate, using the assumptions concerning the random field $(a(x), b(x), \tilde{b}(x))$, that

$$\lim_{\epsilon \downarrow 0} u^\epsilon(x) = Ee^{-\zeta(y)} \int_0^x \exp\{-yE(\frac{b(z)}{a(z)})\}dy,$$

$$\lim_{\epsilon \downarrow 0} v^\epsilon(x) = E\frac{1}{a(y)}e^{\zeta(y)} \int_0^x \exp\{yE(\frac{b(z)}{a(z)})\}dy.$$

Comparing these formulas with the expression for u and v corresponding to a space homogeneous diffusion process, we conclude that the processes X_t^ϵ, $0 \le t \le T$, corresponding to operators

$$L^\epsilon = a(\frac{x}{\epsilon})\frac{d^2}{dx^2} + (b(\frac{x}{\epsilon}) + \frac{1}{\epsilon}\tilde{b}(\frac{x}{\epsilon}))\frac{d}{dx},$$

converge weakly in $C_{0T}(R^1)$ as $\epsilon \downarrow 0$ to the one-dimensional diffusion process corresponding to the operator with constant coefficients

$$\bar{L} = \bar{a}\frac{d^2}{dx^2} + \bar{b}\frac{d}{dx},$$

where

$$\bar{a} = (\phi\psi)^{-1}, \qquad \bar{b} = A(\phi\psi)^{-1}, \quad A = E\left(\frac{b(y)}{a(y)}\right),$$

$$\phi = E\exp\{-\zeta(y)\}, \quad \psi = Ea^{-1}(y)\exp\{\zeta(y)\}.$$

Note that in the self-adjoint case, $L^\epsilon = \frac{d}{dx}(a(\frac{x}{\epsilon})\frac{d}{dx})$, $b(x) = 0$, $\tilde{b}(x) = a'(x)$, $\zeta(x) = \log a(x)$. Thus, $\bar{a} = (Ea^{-1}(y))^{-1}$, $\bar{b} = 0$.

Now let $a(x) \equiv 1$ and $b^\epsilon(x) = \frac{1}{\sqrt{\epsilon}}b(x/\epsilon)$, where $b(x) = b(x,\omega)$ is a mean zero homogeneous random field with sufficiently strong mixing properties. Then a version of the central limit theorem holds, and using the theorems mentioned above we conclude that the stochastic processes X_t^ϵ corresponding to the operators

$$\frac{d^2}{dx^2} + b^\epsilon \frac{d}{dx}$$

converge to the diffusion process with the diffusion coefficient equal to one and with the drift equal to a spatial white noise. Existence of such a process follows from the fact that the functions $u(x)$ and $\vartheta(x)$ defined by (11.4) for $a = 1$ and $b(x)$ equal to white noise are continuous. It is an interesting problem to find a multidimensional analogue of this result.

Consider now another class of homogenization problems. Let $f(x,u)$ be a function of class $\mathcal{F}_1$, 1-periodic in $x \in R^r$ (see §9), and let L be an elliptic operator with 1-periodic coefficients. Consider the wave front propagation problem

$$\frac{\partial u(t,x)}{\partial t} = Lu + f(x,u), \ u(0,x) = 1_G(x); \tag{11.5}$$

$1_G(x)$ is the indicator function of a compact domain $G \subset R^r$ with a sufficiently regular boundary. One can expect that there exists an asymptotic speed of propagation of the area where u is close to 1. The speed can depend on the direction. This problem was examined by Freidlin and Gärtner (see [F6] and the other references there).

Consider, first, the one-dimensional self-adjoint case:

$$\frac{\partial u(t,x)}{\partial t} = \frac{1}{2}\frac{\partial}{\partial x}\left(a(x)\frac{\partial u}{\partial x}\right) + c(x,u)u$$

$$u(0,x) = \begin{cases} 1, & |x| \le 1 \\ 0, & |x| > 1 \end{cases} \tag{11.6}$$

We assume that $a(x)$, $c(x,u)$ are 1-periodic in x, $f(x,u) = c(c,u)u \in \mathcal{F}_1$, $c(x) = c(x,0)$. To formulate the result, introduce an auxiliary eigenvalue problem on the circle T of length 1:

$$L^z\phi(x) = \lambda(z)\phi(x), \ x \in T,$$

where $z \in \mathbb{R}^1$ is a parameter, and

$$L^z \phi = \frac{1}{2}\frac{d}{dx}\left(a(x)\frac{d\phi}{dx}\right) - a(x)z\frac{d\phi}{dx} + \left[c(x) - \frac{1}{2}a'(x)z + \frac{1}{2}a(x)z^2\right]\phi.$$

Let $\lambda(x)$ be the eigenvalue corresponding to the positive eigenfunction. Then, as it is shown in [F6] the asymptotic speed in the positive direction is equal to

$$\vartheta_+^* = \inf_{z>0} \frac{\lambda(z)}{z},$$

and in the negative direction

$$\vartheta_-^* = \inf_{z>0} \frac{\lambda(-z)}{z}.$$

In general $\lambda(z) \neq \lambda(-z)$. If $a(x) = a(1-x)$, $c(x) = c(1-x)$, $\lambda(z) = \lambda(-z)$ and $\vartheta_+^* = \vartheta_-^*$. Thus the "effective" space homogeneous equation should, in general, have a drift term. But in the one-dimensional case it is still possible to write down an "effective" space homogeneous equation (one for different initial functions) for which the front asymptotically behaves like the front in the problem (11.6).

Examine now problem (11.5) for $f(x, u) = c(x, u) \cdot u \in \mathcal{F}_1$ and $r > 1$. Assume that L is self-adjoint. Define

$$L^z \phi = L\phi - (a(x)z \cdot \nabla \phi) + \left[c(x) - \operatorname{div}(a(x)z) + \frac{1}{2}(a(x)z \cdot z)\right]\phi,$$

where $z \in \mathbb{R}^r$, $a(x) = (a^{ij}(x))$. Consider the eigenvalue problem on the r-dimensional torus T:

$$L^z \phi_z(x) = \lambda(z)\phi_z(x), \quad x \in T.$$

Let $\lambda(z)$ be the first eigenvalue of this problem,

$$\vartheta^*(e) = \inf_{z:(e,z)>0} \frac{\lambda(z)}{(e, z)}, \quad e \in \mathbb{R}^r, \ |e| = 1.$$

It is proved that $\vartheta^*(e)$ is the asymptotic speed in the direction e. This means that if we define

$$M = \left\{x \in \mathbb{R}^r : \vartheta^*(\frac{x}{|x|}) \geq |x|\right\}, \qquad (11.7)$$

then

$$\lim_{t \to \infty} u(t, ty) = 0$$

uniformly in y for any compact set F such that $F \cap M = \emptyset$, and

$$\lim_{t \to \infty} u(t, ty) = 1$$

uniformly in y for any compact set belonging to the interior of M.

One can check that M is a convex set containing the origin. But it is not necessary that M is an ellipsoid even if $c(x) = c = constant$. One can derive from the results of §7 that for the operators with constant coefficients and $c(x) = c$ the set M is always an ellipsoid. Thus in the multidimensional case the "effective" space homogeneous equation does not exist.

One can consider the wave front propagation problem for equations with slowly changing periodic coefficients: $a^{ij}(x) = a^{ij}(x, \frac{x}{\epsilon})$, where $a^{ij}(x, y)$ are smooth functions 1-periodic in the second variable; $c(x) = c = const$:

$$\frac{\partial u^\epsilon(t, x)}{\partial t} = \frac{\epsilon}{2} \sum_{i,j=1}^{r} a^{ij}(x, \frac{x}{\epsilon}) \frac{\partial^2 u^\epsilon}{\partial x^i \, \partial x^j} + \frac{1}{\epsilon} f(u^\epsilon),$$

$$t > 0, \ x \in \mathbf{R}^r, \ u^\epsilon(0, x) = \chi_{G_0}(x),$$

where $\chi_{G_0}(x)$ is the indicator function of $G_0 \in \mathbf{R}^r$; $f(\cdot) \in \mathcal{F}_1$, $f'(0) = c$. In this case the motion of the front can be described by Huygens principle such that the corresponding velocity field is isotropic and homogeneous if calculated in a Finsler metric. This Finsler metric is defined by its unit spheres in the tangent space at each $w \in \mathbf{R}^r$. The sphere at the point $w \in \mathbf{R}^r$ is defined by (11.7) where $\vartheta^*(e)$ is calculated for the coefficients $a_w^{ij}(x) = a^{ij}(w, x)$.

Consider now a two-parameter homogenization problem:

$$\frac{\partial u^{\epsilon,\delta}(t, x)}{\partial t} = \frac{\delta}{2} \sum_{i,j=1}^{r} a^{ij}(\frac{x}{\epsilon}) \frac{\partial^2 u^{\epsilon,\delta}}{\partial x^i \partial x^j} + \frac{1}{\delta} f(\frac{x}{\epsilon}, u^{\epsilon,\delta}),$$

$$t > 0, \ x \in \mathbf{R}^r, \ u^{\epsilon,\delta}(0, x) = \chi_{G_0}(x).$$

We assume that $f(x, \cdot) \in \mathcal{F}_1$, $a^{ij}(x)$ and $f(x, u)$ are smooth and 1-periodic in x, $\sum_{i,j=1}^{r} a^{ij}(x) \lambda_i \lambda_j \geq \bar{a} |\lambda|^2$, $0 < \epsilon, \delta \ll 1$. In this case the function $u^{\epsilon,\delta}(t, x)$ also converges to a step-function with values 0 and 1 as $\epsilon, \delta \downarrow 0$. The motion of the front (interface) can be also described by a Huygens principle. But the velocity field will be different for different relations between ϵ and δ.

If $\epsilon \delta^{-1}$ tends to zero then the averaging occurs first: Let $m(x)$ be the invariant density for the process on 1-torus T governed by the operator:

$$L = \frac{1}{2} \sum_{i,j=1}^{r} a^{ij}(x) \frac{\partial^2}{\partial x^i \partial x^j}; \quad \bar{a}^{ij} = \int_T a^{ij}(x) m(x) \, dx,$$

$$\bar{f}(u) = \int_T f(x, u) m(x) \, dx, \quad (\bar{a}_{ij}) = (\bar{a}^{ij})^{-1}.$$

Denote by $\bar{\rho}(\cdot, \cdot)$ the Riemannian metric corresponding to the form $\sum \bar{a}_{ij} \, dx^i \, dx^j$. Then the motion of the interface is governed by the Huygens principle such that the corresponding velocity field $\vartheta(x, e)$, $x \in \mathbb{R}^r$, $e \in \mathbb{R}^r$, $|e| = 1$, calculated in the metric $\bar{\rho}$ is isotropic homogeneous and $|\vartheta(x, e)| = \sqrt{2f'(0)}$. This follows from the fact that if $0 < \epsilon \ll \delta \ll 1$, then the rate of convergence of the corresponding diffusion process on the torus to the invariant distribution is better than $\exp\{-\frac{A}{\delta}\}$ for any $A > 0$.

If ϵ and δ have the same order the motion of the front is governed by the Huygens principle with the velocity field homogeneous and isotropic in the Finsler metric described above (see [GF2], [F6]).

In the case when $\epsilon \delta^{-1}$ tends to zero the motion of the front is also governed by the Huygens principle. The corresponding velocity field is homogeneous and isotropic in a Finsler metric which can be described as follows: for any $\epsilon > 0$ denote by $\rho^\epsilon(\cdot, \cdot)$ the Riemannian metric corresponding to the form $\sum_{i,j=1}^r a_{ij}(\frac{x}{\epsilon}) dx^i dx^j$. Using the Kingman subadditive ergodic theorem, one can prove that $\lim_{\epsilon \downarrow 0} \rho^\epsilon(x, y) = \hat{\rho}(x, y)$ exists; $\hat{\rho}(\cdot, \cdot)$ is the Finsler metric in which the velocity field is isotropic and homogeneous.

In the conclusion of this section, consider the homogenization problem for linear RDE-systems:

$$\frac{\partial u_k^{\epsilon,\delta}(t, x)}{\partial t} = \frac{1}{2} \sum_{i,j=1}^r a_k^{ij}(x/\epsilon) \frac{\partial^2 u_k^{\epsilon,\delta}(t, x)}{\partial x^i \partial x^j} + \sum_{i=1}^r b_k^i(x/\epsilon) \frac{\partial u_k^{\epsilon,\delta}(t, x)}{\partial x^i} +$$

(11.8)

$$+ \frac{1}{\delta} \sum_{j=1}^n d_{kj}(x/\epsilon)(u_j^{\epsilon,\delta}(t, x) - u_k^{\epsilon,\delta}(t, x)),$$

$$t > 0, \ x \in \mathbb{R}^r, \ k \in \{1, \ldots, n\}; \epsilon, \delta > 0.$$

We assume that the coefficients $a_k^{ij}(x)$, $b_k^i(x)$, $d_{kj}(x)$ are smooth enough and 1-periodic in each $x^i, 1 \leq i \leq r; d_{kj}(x) > 0, \sum_{i,j=1}^r a_k^{ij}(x)\lambda_i\lambda_j \geq \bar{a} > 0$.

Denote by $(X_t^{\epsilon,\delta}, \nu_t^{\epsilon,\delta})$ the Markov process in $\mathbb{R}^r \times \{1, \ldots, n\}$ corresponding to (11.8). The limiting behavior of this process depends on the relationship between ϵ and δ. First, let $\delta = 1$, $\epsilon \downarrow 0$. Then, roughly speaking, the averaging in x occurs before any particle changes its type. Let $m_k(x)$ be the normalized invariant measure of the process on the 1-torus T governed by the "shortened" k-th operator

$$\hat{L}_k = \frac{1}{2} \sum_{i,j=1}^r a_k^{ij}(x) \frac{\partial^2}{\partial x^i \partial x^j}.$$

Note that $m_k(x)$ is the unique solution of the problem

$$\hat{L}_k^* m_k(x) = \frac{1}{2} \sum_{i,j=1}^r \frac{\partial^2}{\partial x^i \partial x^j}(a_k^{ij}(x)m_k(x)) = 0, \ \int_T m_k(x)dx = 1.$$

Define $R_k[f] = \int_T f(x)m_k(x)dx$, for any measurable $f(x)$, $x \in T$; $\bar{a}_k^{ij} = R_k[a_k^{ij}]$, $\bar{b}_k^i = R_k[b_k^i]$, $\bar{d}_{kj} = R_k[d_{kj}]$. The processes $(X_t^{\epsilon,1}, \nu_t^{\epsilon,1})$, $0 \le t \le T$, converge weakly as $\epsilon \downarrow 0$ to the process governed by the system of (11.8) type, but with $a_k^{ij}(x)$, $b_k^i(x)$, $d_{kj}(x)$ replaced by $\bar{a}_k^{ij}(x)$, $\bar{b}_k^i(x)$, $\bar{d}_{kj}(x)$.

Now, if $\delta \downarrow 0, \delta \gg \epsilon^2$, the future homogenization is possible. Let $(\bar{q}_1, \ldots, \bar{q}_n)$ be the invariant distribution for the continuous time Markov chain on $\{1, \ldots, n\}$ with the transition intensities $\bar{d}_{kj}$. Then $X_t^{\epsilon,\delta}, 0 \le t \le T$, converges weakly in the space of continuous functions as ϵ, $\delta \downarrow 0$, $\epsilon^2 \ll \delta$, to the Markov Gaussian process in R^r governed by the operator

$$\bar{L} = \frac{1}{2}\sum_{i,j}(\sum_k \bar{a}_k^{ij}\bar{q}_k)\frac{\partial^2}{\partial x^i \partial x^j} + \sum_i(\sum_k \bar{b}_k^i\bar{q}_k)\frac{\partial}{\partial x^i}.$$

Here we have homogenization not just in space, but in the type of the particle as well: one can introduce an "effective" particle governed by $\bar{L}$.

Now let $\delta = \epsilon^2$. Then the mixing in space and type have the same rate and the averaging should be made in both components simultaneously. Define $m_k(x)dx$, $x \in T$, $k \in \{1, \ldots, n\}$, as the invariant measure for the process on $T \times \{1, \ldots, n\}$ governed by the system (11.8) for $\epsilon = \delta = 1$. Let

$$\hat{a}^{ij} = \sum_{k=1}^n \int_T a_k^{ij}(x)m_k(x)\,dx, \qquad \hat{b}^i = \sum_{k=1}^n \int_T b_k^i(x)m_k(x)\,dx.$$

Then the process $X_t^{\epsilon,\epsilon^2}$ converges as $\epsilon \downarrow 0$ to the process corresponding to the operator

$$\hat{L} = \frac{1}{2}\sum_{i,j=1}^r \hat{a}^{ij}\frac{\partial^2}{\partial x^i \partial x^j} + \sum_{i=1}^r \hat{b}^i\frac{\partial}{\partial x^i}.$$

Using the representations of the solutions of various initial-boundary problems for system (11.8), one can derive from these statements results concerning the behavior of these solutions as $\epsilon, \delta \to 0$.

Assume now that $\epsilon, \delta \to 0$, $\delta \ll \epsilon^2$. Then the averaging of the type of the particle goes first. If $\{\tilde{q}_r(x)\}$ is the invariant distribution for the continuous time Markov chain in $\{1, \ldots, n\}$ with the transition intensities $d_{kj}(x), x$ is a parameter, and $\tilde{m}(x)$ is the invariant density for the process on the torus T governed by the operator

$$\tilde{L} = \frac{1}{2}\sum_{i,j}(\sum_k a_k^{ij}\tilde{q}_k(x))\frac{\partial^2}{\partial x^i \partial x^j},$$

then the processes $X_t^{\epsilon,\delta}$ converge as $\epsilon, \delta \to 0$, $\delta \ll \epsilon^2$, to the process in R^r governed by the operator

$$\bar{\bar{L}} = \frac{1}{2} \sum_{i,j=1}^r \bar{\bar{a}}^{ij} \frac{\partial^2}{\partial x^i \partial x^j} + \sum_{i=1}^r \bar{\bar{b}}^i \frac{\partial}{\partial x^i},$$

$$\bar{\bar{a}}^{ij} = \sum_{k=1}^n \int_T a_k^{ij}(x)\tilde{q}_k(x)m_k(x)\,dx, \qquad \bar{\bar{b}}^i(x) = \sum_{k=1}^n \int_T b_k^i(x)\tilde{q}_k(x)m_k(x)\,dx.$$

One can see that the operators $\bar{\bar{L}}$ and $\hat{L}$ are, in general, different. This means that the homogenization in space and in type of the particles are not commutative. It is actually a general property: homogenization in different components are not commutative. For example, consider the process $X_t^{\epsilon,\delta}$ in R^2 corresponding to the operator

$$L^{\epsilon,\delta} = A(\frac{x}{\epsilon}, \frac{y}{\delta})\frac{\partial^2}{\partial x^2} + B(\frac{x}{\epsilon}, \frac{y}{\delta})\frac{\partial^2}{\partial y^2})$$

where $A(x,y)$ and $B(x,y)$ are smooth positive functions 1-periodic in x and y. The the "effective" operators with constant coefficients as $\epsilon, \delta \downarrow 0$ can be different for the cases $\epsilon \ll \delta$ and $\delta \ll \epsilon$.

Finally, consider the operators (1.2) with $\epsilon = 1$ and $\delta \downarrow 0$. Then processes $X_t^{\epsilon,\delta} = X_t^\delta$ converge as $\delta \downarrow 0$ to a diffusion process in R^r governed by the averaged operator $\bar{\bar{L}} : \bar{\bar{L}} = \sum_{k=1}^n q_k(x)L_k$. Here $L_k = \frac{1}{2} \sum_{i,j=1}^r a_k^{ij}(x)\frac{\partial^2}{\partial x^i \partial x^j} + \sum b_k^i(x)\frac{\partial}{\partial x^i}$, and $\{q_k(x)\}$ is the invariant distribution for a continuous time Markov chain in $\{1,\ldots,n\}$ with transition intensities $d_{ij}(x)$, x is a parameter. Such a distribution exists and is unique if all $d_{ij}(x) > 0$. If for some x the intensities are equal to zero, the limiting averaged process can have a more complicated phase space. For example, let $x \in R^1, d_{ij}(x) > 0$ for $x \geq 0$, and $d_{ij} \equiv 0$ for $x < 0$. The processes X_t^δ converges weakly to a diffusion process on a graph Γ, consisting of one vertex 0 and $n+1$ half-lines connected to 0 : $I_0, I_1, \ldots, I_n$. The half-line I_0 corresponds to $x > 0$. The process on I_0 is defined by the operator $\bar{\bar{L}}$ described above. On each $I_k, k = 1, \ldots, n$, the process is governed by the operator L_k. As we have seen in Section 5, to define the process on Γ, one needs to describe its behavior at the vertex 0. This behavior is defined by gluing conditions for functions $f(x), x \in \Gamma$, belonging to the domain of the generator of the process. $f(x)$ is continuous on $\Gamma, \lim_{x \to 0, x \in I_k} L_k f(x) = Lf(x)$ is independent of $k \in \{1, \ldots, n\}$, and

$$\frac{df(0)}{dx_0} = \sum_{k=1}^n q_k \frac{df(0)}{dx_k}.$$

Here x_k is the distance of the point $x_k \in I_k$ from 0, $k = 0, 1, \ldots, n$; $q_k = \lim_{x \to 0} q_k(x)$. This problem is considered in [FW2].

References

[A]   V. I. Arnold, (1991), Topological and ergodic properties of closed 1-forms with incommensurable period, Functional Analysis and Appl. 23, 2, 1–12.

[Az]   R. Azencott (1978), Grandes déviations et applications, Lecture Notes in Math., 774, Springer.

[BeF] G. BenArous and M. Freidlin (1994), Markov processes and RDEs, In preparation.

[BLP] A. Bensoussan, J. J. Lions, G. C. Papanicolaou (1978), Asymptotic Analysis for periodic Structures, North-Holland Publ. Company, Amsterdam-New York-Oxford.

[BlF] Yu. N. Blagoveschenskii and M. Freidlin (1961), Some properties of diffusion processes depending on a parameter, Soviet Math. Dokl., 138, 508–511.

[BF]   A. N. Borodin and M. Freidlin (1994), Fast oscillating random perturbations of dynamical systems with conservation laws, Ann. Inst. H. Poincaré (to appear).

[BM]   N. M. Bogolyubov and Y. A. Mitropolskii (1961), Asymptotic Methods in the Theory of Nonlinear Oscillations, $2^{\underline{nd}}$ed., Gordon & Breach, New York, Delhi.

[DS]   J.-D. Deuschel and D. W. Stroock (1989), Large Deviations, Academic Press, Boston etc.

[DF]   J. P. Dunyak and M. I. Freidlin (1993), Optimal stabilization of Hamiltonian systems perturbed by white noise, Preprint.

[D]   E. B. Dynkin, Markov Processes (1965), Springer, Berlin etc.

[EF1] A. Eizenberg and M. Freidlin (1990), On the Dirichlet problem for a class of PDE systems with small parameter, Stochastics and Stoch. Reports, 33, 111–148.

[EF2] A. Eizenberg and M. Freidlin (1993), Large deviations for Markov processes corresponding to PDE systems, Annals of Probab., 21, 2, 1015–1044.

[EF3] A. Eizenberg and M. Freidlin (1993), Averaging principle for random evolution equation and corresponding Dirichlet problems, Probab. Theory Related Fields, 94, 335–374.

[EK]   S. N. Ethier and T. G. Kurtz (1986), Markov Processes, John Wiley and Sons, New York.

[Fe] W. Feller (1957), Generalized second order differential operators and their lateral conditions, Illinois J. of Math., 1, 459–504.

[Fi] R. A. Fisher (1937), The wave of advance of advantageous genes, Annals of Eugenics, 7, 355–369.

[F1] M. I. Freidlin (1964), Dirichlet problem for equations with periodic coefficients, Probab. Theory and Appl., 9, 133–139.

[F2] M. I. Freidlin (1976), Fluctuations in dynamical systems with averaging, Soviet Math. Dokl. 17, 1, 104–108.

[F3] M. I. Freidlin (1977), Sublimiting distributions and stabilization of solutions of parabolic equations with a small parameter, Soviet Math. Dokl. 18, 44, 114–118.

[F4] M. I. Freidlin (1979), Propagation of concentration waves due to a random motion connected with growth, Soviet Math. Dokl. 246, 544–548.

[F5] M. I. Freidlin (1981), On elliptic equations with a small parameter, C.R. Math. Acad. Sci. Canada, 111, 4, 209–214.

[F6] M. I. Freidlin (1985), Functional Integration and Partial Differential Equations, Princeton Univ. Press.

[F7] M. I. Freidlin (1985), Limit theorems for large deviations and reaction diffusion equations, Ann. of Probab., 13, 3, 639–676.

[F8] M. I. Freidlin (1991), Coupled reaction-diffusion equations, Annals of Probab., 19, 1, 29–57.

[F9] M. I. Freidlin (1992), Semi-linear PDEs and Limit Theorem for Large Deviations, Lectures in Saint-Flour Summer School in Probality, Lecture Notes in Math., 1527, Springer.

[F10] M. I. Freidlin (1995), Wave Fronts Propagation for KPP-type Equations, Surveys in Appl. Math., Volume 2, J. B. Keller, D. W. McLaughlin and G. Papanicolaou (Editors), pp. 1–62. Plenum

[F11] M. I. Freidlin (1994), Random Perturbations of Dynamical Systems: Large Deviations and Averaging, Math. Journ. Univ. of San Paulo, 1, 2/3, 183–216.

[FL1] M. I. Freidlin and Tzong-Yow Lee (1992), Large deviation principle for diffusion-transmutation processes and Dirichlet problem for PDE systems with a small parameter, Preprint.

[FL2] M. I. Freidlin and Tzong-Yow Lee (1992), Wave front propagation and large deviations for diffusion-transmutation processes, Preprint.

[FW1] M. I. Freidlin and A. D. Wentzell (1984), Random Perturbations of Dynamical Systems, Springer (translation from the Russian, Nauka 1979).

[FW2] M. I. Freidlin and A. D. Wentzell (1993), Diffusion processes on graphs and averaging principle, Annals of Probab., 21, 4, 2215–2245.

[FW3] M. I. Freidlin and A. D. Wentzell (1994), Random Perturbations of Hamiltonian Systems, Memoirs of AMS.

[FW4] M. I. Freidlin and A. D. Wentzell (1994), Necessary and sufficient conditions for weak convergence of one-dimensional Markov processes. Festschrift dedicated to 70-th birthday of Professor E.B. Dynkin, Birkhäuser, M. I. Freidlin (Editor).

[G] Ju. Gärtner (1980), Nonlinear diffusion equations and excitable media, Soviet Math. Dokl. 254, 1310–1314.

[GF1] Ju. Gärtner and M. I. Freidlin (1978), A new contribution to large deviations for stochastic processes, Vestnik Mosc. Univ., Ser. Math., 5, 52–59 (in Russian).

[GF2] Ju. Gärtner and M. I. Freidlin (1979), Propagation of concentration waves in periodic and random media, Soviet Math. Dokl., 249, 521–525.

[Kh1] R. Z. Khasminskii (1966), A limit theorem for solutions of differential equations with random right-hand side, Theory of Probab. Appl., 11, 3, 390–406.

[Kh2] R. Z. Khasminskii (1966), On processes defined by differential equations with a small parameter, Theory of Probab. Appl., 11, 2, 211–228.

[KPP] A. Kolmogorov, I. Petrovskii and N. Piskunov (1937), Étude de l'èquation de la diffusion avec croissence de la matière et son application a un problème biologique, Moscov Univ. Bull. Math., 1, 1–25.

[K] S. Kozlov (1979), The Averaging of Random Operators, Math. USSR Sbornik, 37,2,167–180.

[PV] G. Papanicolaou and S. R. S. Varadhan (1981), Boundary value problems with rapidly oscillating random coefficients, Proceedings of the Conference on Random Fields, Esztergon, Hungary, Colloquia Math. Soc. Janos Bolyai, 27, 835–873.

[R] H. Rund (1959), The Differential Geometry of Finsler Spaces, Springer Verlag, Berlin.

[S]   R. L. Stratonovich (1966), Conditional Markov Processes, Moscow Univ. Press (Russian).

[SKh] Ya. G. Sinai and K. M. Khanin (1992), Mixing for some classes of special flows over a circle rotation, Functional Analysis and Appl., 26, 3, 1–21.

[SV]   D. W. Stroock and S. R. S. Varadhan (1979), Multidimensional Diffusion Processes, Springer Verlag, Berlin.

[V1]   S. R. S. Varadhan (1967), Diffusion processes in a small time interval, Comm. Pure and Appl. Math., 20:4, 659–685.

[V2]   S. R. S. Varadhan (1984), Large Deviations and Applications, SIAM, Philadelphia.

[ZKO] V. V. Zhikov, S. M. Kozlov, O. A. Oleinik and Ha Ten Ngoan (1979), Averaging and G-convergence of differentiable operators, Russian Math. Surveys, 34:5, 69–147.

Index

LM –
Lectures in Mathematics, ETH Zürich

Department of Mathematics
Research Institute of Mathematics

*Each year the Eidgenössische Technische Hochschule (ETH) at Zürich invites a selected group of mathematicians to give postgraduate seminars in various areas of pure and applied mathematics. These seminars are directed to an audience of many levels and backgrounds. Now some of the most successful lectures are being published for a wider audience through the **Lectures in Mathematics, ETH Zürich** series. Lively and informal in style, moderate in size and price, these books will appeal to professionals and students alike, bringing a quick understanding of some important areas of current research.*

C. de Boor
Splinefunktionen
1990. ISBN 3-7643-2514-3

J.D. Monk
Cardinal Functions on Boolean Algebras
1990. ISBN 3-7643-2495-3

D. Bättig/H. Knörrer
Singularitäten
1991. ISBN 3-7643-2616-6

R.J. LeVeque
Numerical Methods for Conservation Laws
2nd Edition, 3rd Printing 1994.
1992. ISBN 3-7643-2723-5

R. Narasimhan
Compact Riemann Surfaces
1992. ISBN 3-7643-2742-1

A.J. Tromba
Teichmüller Theory in Riemannian Geometry
1992. ISBN 3-7643-2735-9

M. Yor
Some Aspects of Brownian Motion
1992. ISBN 3-7643-2807-X

G. Baumslag
Topics in Combinatorial Group Theory
1993. ISBN 3-7643-2921-1

M. Giaquinta
Introduction to Regularity Theory for Nonlinear Elliptic Systems
1993. ISBN 3-7643-2879-7

O. Nevanlinna
Convergence of Iterations for Linear Equations
1993. ISBN 3-7643-2865-7

R.-P. Holzapfel
The Ball and Some Hilbert Problems
1995. ISBN 3-7643-2835-5

J.F. Carlson
Modules and Group Algebras
Notes by Ruedi Suter
1996. ISBN 3-7643-5389-9